UAV PILOT LOGBOOK
PRO

HOLDER'S NAME:

Logbook number _____
Entries from _____
Through _____

Copyright © 2018 by **Parhelion Aerospace GmbH**

www.parhelionaerospace.com
Author: Michael L. Rampey
ISBN: 978-2-8399-2433-7

All rights reserved. Reproduction in any manner, without written permission, is prohibited.

HOLDER

Holder's Operator Certificate	Holder's 2nd Operator Certificate
Certificate	Certificate
Number	Number
Certification authority	Certification authority
Date / place of initial qualification	Date / place of initial qualification
Holder's address	*(Space for address change)*

CONTENTS

Topic	Page No.	Topic	Page No.
Title page, Copyright page, Holder page and Contents	1-4	Flight Log Pages	26
Part 1: Aircraft and Equipment Data	5	Per-aircraft Totals	276
Aircraft Identification Data	6	**Part 3: Record of Pilot Training**	277
Aircraft Batteries Data	10	Flight and Ground Training Record	278
Controller Batteries Data	14	Training Certifications	280
Other Power Sources Data	16	Additional Notes	285
Equipment	18		
Part 2: Pilot Flight Log	21		
Notes on the Recording of Pilot Flight Data	22		
Flight and Technical Data Examples Pages	24		

Notes on the use of this book:

Use Part 1 to record aircraft and equipment characteristics.

Use Part 2 to record all flights.

Use Part 3 to record flight and ground training courses taken and certification of training programs completed.

Part 1: Aircraft and Equipment Data

Aircraft Identification Data

1. Make _____ Model _____ Year of Manufacture _____

   Registration Number _____ Type (e.g. Fixed-wing, Rotary, LTA) _____ Date Acquired _____

   Owner's Name / Address _____

2. Make _____ Model _____ Year of Manufacture _____

   Registration Number _____ Type (e.g. Fixed-wing, Rotary, LTA) _____ Date Acquired _____

   Owner's Name / Address _____

3. Make _____ Model _____ Year of Manufacture _____

   Registration Number _____ Type (e.g. Fixed-wing, Rotary, LTA) _____ Date Acquired _____

   Owner's Name / Address _____

4. Make _____ Model _____ Year of Manufacture _____

   Registration Number _____ Type (e.g. Fixed-wing, Rotary, LTA) _____ Date Acquired _____

   Owner's Name / Address _____

Aircraft Identification Data, continued

5. Make _____ Model _____ Year of Manufacture _____

Registration Number _____ Type (e.g. Fixed-wing, Rotary, LTA) _____ Date Acquired _____

Owner's Name / Address _____

6. Make _____ Model _____ Year of Manufacture _____

Registration Number _____ Type (e.g. Fixed-wing, Rotary, LTA) _____ Date Acquired _____

Owner's Name / Address _____

7. Make _____ Model _____ Year of Manufacture _____

Registration Number _____ Type (e.g. Fixed-wing, Rotary, LTA) _____ Date Acquired _____

Owner's Name / Address _____

8. Make _____ Model _____ Year of Manufacture _____

Registration Number _____ Type (e.g. Fixed-wing, Rotary, LTA) _____ Date Acquired _____

Owner's Name / Address _____

AIRCRAFT IDENTIFICATION DATA, CONTINUED

9. Make _____ Model _____ Year of Manufacture _____

Registration Number _____ Type (e.g. Fixed-wing, Rotary, LTA) _____ Date Acquired _____

Owner's Name / Address _____

10. Make _____ Model _____ Year of Manufacture _____

Registration Number _____ Type (e.g. Fixed-wing, Rotary, LTA) _____ Date Acquired _____

Owner's Name / Address _____

11. Make _____ Model _____ Year of Manufacture _____

Registration Number _____ Type (e.g. Fixed-wing, Rotary, LTA) _____ Date Acquired _____

Owner's Name / Address _____

12. Make _____ Model _____ Year of Manufacture _____

Registration Number _____ Type (e.g. Fixed-wing, Rotary, LTA) _____ Date Acquired _____

Owner's Name / Address _____

AIRCRAFT IDENTIFICATION DATA, CONTINUED

13. Make   Model   Year of Manufacture

Registration Number   Type (e.g. Fixed-wing, Rotary, LTA)   Date Acquired

Owner's Name / Address

14. Make   Model   Year of Manufacture

Registration Number   Type (e.g. Fixed-wing, Rotary, LTA)   Date Acquired

Owner's Name / Address

15. Make   Model   Year of Manufacture

Registration Number   Type (e.g. Fixed-wing, Rotary, LTA)   Date Acquired

Owner's Name / Address

16. Make   Model   Year of Manufacture

Registration Number   Type (e.g. Fixed-wing, Rotary, LTA)   Date Acquired

Owner's Name / Address

Aircraft Batteries Data

("Out," with checkbox, designates a battery no longer in use)

Batteries for (aircraft designation): _____

Battery Designation	Out:	Out:	Out:	Out:	Out:	Out:
Make / Type						
Date Acquired						
mAh / Vmax						

Batteries for (aircraft designation): _____

Battery Designation	Out:	Out:	Out:	Out:	Out:	Out:
Make / Type						
Date Acquired						
mAh / Vmax						

Batteries for (aircraft designation): _____

Battery Designation	Out:	Out:	Out:	Out:	Out:	Out:
Make / Type						
Date Acquired						
mAh / Vmax						

Batteries for (aircraft designation): _____

Battery Designation	Out:	Out:	Out:	Out:	Out:	Out:
Make / Type						
Date Acquired						
mAh / Vmax						

Aircraft Batteries Data, continued

("Out," with checkbox, designates a battery no longer in use)

Batteries for (aircraft designation): _____

Battery Designation	Out: ☐	Out: ☐	Out: ☐	Out: ☐
Make / Type				
Date Acquired				
mAh / Vmax				

Batteries for (aircraft designation): _____

Battery Designation	Out: ☐	Out: ☐	Out: ☐	Out: ☐
Make / Type				
Date Acquired				
mAh / Vmax				

Batteries for (aircraft designation): _____

Battery Designation	Out: ☐	Out: ☐	Out: ☐	Out: ☐
Make / Type				
Date Acquired				
mAh / Vmax				

Batteries for (aircraft designation): _____

Battery Designation	Out: ☐	Out: ☐	Out: ☐	Out: ☐
Make / Type				
Date Acquired				
mAh / Vmax				

Batteries for (aircraft designation): _____

Battery Designation	Out: ☐	Out: ☐	Out: ☐	Out: ☐
Make / Type				
Date Acquired				
mAh / Vmax				

Batteries for (aircraft designation): _____

Battery Designation	Out: ☐	Out: ☐	Out: ☐	Out: ☐
Make / Type				
Date Acquired				
mAh / Vmax				

Aircraft Batteries Data, continued

("Out," with checkbox, designates a battery no longer in use)

Batteries for (aircraft designation): _____

Battery Designation	Out: ☐	Out: ☐	Out: ☐	Out: ☐	Out: ☐	Out: ☐
Make / Type						
Date Acquired						
mAh / Vmax						

Batteries for (aircraft designation): _____

Battery Designation	Out: ☐	Out: ☐	Out: ☐	Out: ☐	Out: ☐	Out: ☐
Make / Type						
Date Acquired						
mAh / Vmax						

Batteries for (aircraft designation): _____

Battery Designation	Out: ☐	Out: ☐	Out: ☐	Out: ☐	Out: ☐	Out: ☐
Make / Type						
Date Acquired						
mAh / Vmax						

Batteries for (aircraft designation): _____

Battery Designation	Out: ☐	Out: ☐	Out: ☐	Out: ☐	Out: ☐	Out: ☐
Make / Type						
Date Acquired						
mAh / Vmax						

Aircraft Batteries Data, continued

("Out," with checkbox, designates a battery no longer in use)

Batteries for (aircraft designation): _____

Battery Designation	Out: ☐	Out: ☐	Out: ☐	Out: ☐	Out: ☐
Make / Type					
Date Acquired					
mAh / Vmax					

Batteries for (aircraft designation): _____

Battery Designation	Out: ☐	Out: ☐	Out: ☐	Out: ☐	Out: ☐
Make / Type					
Date Acquired					
mAh / Vmax					

Batteries for (aircraft designation): _____

Battery Designation	Out: ☐	Out: ☐	Out: ☐	Out: ☐	Out: ☐
Make / Type					
Date Acquired					
mAh / Vmax					

Batteries for (aircraft designation): _____

Battery Designation	Out: ☐	Out: ☐	Out: ☐	Out: ☐	Out: ☐
Make / Type					
Date Acquired					
mAh / Vmax					

Batteries for (aircraft designation): _____

Battery Designation	Out: ☐	Out: ☐	Out: ☐	Out: ☐	Out: ☐
Make / Type					
Date Acquired					
mAh / Vmax					

Batteries for (aircraft designation): _____

Battery Designation	Out: ☐	Out: ☐	Out: ☐	Out: ☐	Out: ☐
Make / Type					
Date Acquired					
mAh / Vmax					

Controller Batteries Data

("Out," with checkbox, designates a battery no longer in use)

Batteries for (controller designation): _____

	Out: ☐	Out: ☐	Out: ☐	Out: ☐	Out: ☐
Battery Designation					
Make / Type					
Date Acquired					
mAh / Vmax					

Batteries for (controller designation): _____

	Out: ☐	Out: ☐	Out: ☐	Out: ☐	Out: ☐
Battery Designation					
Make / Type					
Date Acquired					
mAh / Vmax					

Batteries for (controller designation): _____

	Out: ☐	Out: ☐	Out: ☐	Out: ☐	Out: ☐
Battery Designation					
Make / Type					
Date Acquired					
mAh / Vmax					

Batteries for (controller designation): _____

	Out: ☐	Out: ☐	Out: ☐	Out: ☐	Out: ☐
Battery Designation					
Make / Type					
Date Acquired					
mAh / Vmax					

Batteries for (controller designation): _____

	Out: ☐	Out: ☐	Out: ☐	Out: ☐	Out: ☐
Battery Designation					
Make / Type					
Date Acquired					
mAh / Vmax					

Batteries for (controller designation): _____

	Out: ☐	Out: ☐	Out: ☐	Out: ☐	Out: ☐
Battery Designation					
Make / Type					
Date Acquired					
mAh / Vmax					

Controller Batteries Data, continued

("Out," with checkbox, designates a battery no longer in use)

Batteries for (controller designation): _____

Battery Designation	Out: ☐	Out: ☐	Out: ☐	Out: ☐
Make / Type				
Date Acquired				
mAh / Vmax				

Batteries for (controller designation): _____

Battery Designation	Out: ☐	Out: ☐	Out: ☐	Out: ☐
Make / Type				
Date Acquired				
mAh / Vmax				

Batteries for (controller designation): _____

Battery Designation	Out: ☐	Out: ☐	Out: ☐	Out: ☐
Make / Type				
Date Acquired				
mAh / Vmax				

Batteries for (controller designation): _____

Battery Designation	Out: ☐	Out: ☐	Out: ☐	Out: ☐
Make / Type				
Date Acquired				
mAh / Vmax				

Batteries for (controller designation): _____

Battery Designation	Out: ☐	Out: ☐	Out: ☐	Out: ☐
Make / Type				
Date Acquired				
mAh / Vmax				

Batteries for (controller designation): _____

Battery Designation	Out: ☐	Out: ☐	Out: ☐	Out: ☐
Make / Type				
Date Acquired				
mAh / Vmax				

Other Power Sources Data

Liquid fuel for (aircraft designation): _____	Buoyancy Gas for (aircraft designation): _____
Type	Type
Full Tank Quantity	Full Aircraft Quantity
Notes:	Notes
Liquid fuel for (aircraft designation): _____	Buoyancy Gas for (aircraft designation): _____
Type	Type
Full Tank Quantity	Full Aircraft Quantity
Notes:	Notes
Other Liquid fuel Data	Other Buoyancy Gas Data

Other Power Sources Data, continued

Liquid fuel for (aircraft designation): _____	Buoyancy Gas for (aircraft designation): _____
Type	Type
Full Tank Quantity	Full Aircraft Quantity
Notes:	Notes
Liquid fuel for (aircraft designation): _____	Buoyancy Gas for (aircraft designation): _____
Type	Type
Full Tank Quantity	Full Aircraft Quantity
Notes:	Notes
Other Liquid fuel Data	Other Buoyancy Gas Data

EQUIPMENT

("Out," with checkbox, designates a device no longer in use)

	Make / Model / Identification Number / Purpose of equipment recorded in this book (e.g. cameras, lenses, sensors)	Date Acquired
1.		Out: ☐
2.		Out: ☐
3.		Out: ☐
4.		Out: ☐
5.		Out: ☐
6.		Out: ☐
7.		Out: ☐
8.		Out: ☐
9.		Out: ☐
10.		Out: ☐

Equipment, continued

("Out," with checkbox, designates a device no longer in use)

Make / Model / Identification Number / Purpose of equipment recorded in this book (e.g. cameras, lenses, sensors)	Date Acquired
11.	Out: ☐
12.	Out: ☐
13.	Out: ☐
14.	Out: ☐
15.	Out: ☐
16.	Out: ☐
17.	Out: ☐
18.	Out: ☐
19.	Out: ☐
20.	Out: ☐

Equipment, continued

("Out," with checkbox, designates a device no longer in use)

Make / Model / Identification Number / Purpose of equipment recorded in this book (e.g. cameras, lenses, sensors)	Date Acquired
21. ☐ Out:	
22. ☐ Out:	
23. ☐ Out:	
24. ☐ Out:	
25. ☐ Out:	
26. ☐ Out:	
27. ☐ Out:	
28. ☐ Out:	
29. ☐ Out:	
30. ☐ Out:	

Part 2: Pilot Flight Log

Notes on the recording of pilot flight data (on the even-numbered pages, beginning with page 26) in this book:

For purposes of logging flight data, a "flight" might be conveniently defined as lasting from initial takeoff to the final landing that is followed by the motor(s) being switched off. In other words, some flights might have several landings, but if the motors are not switched off and the aircraft is immediately re-launched the several flight segments, from launch to final landing/motors off, could be counted as one flight.

By columns across the top of each logging page / per flight:

Flt. No. Enter a sequential number for each flight (e.g. 1, 2, 3, or F1, F2, F3, etc.).

Date Enter the current year in the space provided at the upper left-hand corner of each flight data page. Enter the day and month on which each flight commences in the spaces given.

Aircraft Enter the make, model and registration number of the aircraft flown.

Flight Location Enter the takeoff location and time and the landing location and time. Note that the takeoff and landing may occur at the same place and need be entered only once for both phases of flight. Remark whether you use local time or UTC.

Flight Duration Enter the duration of each flight, in hours and minutes or hours and tenths, in the column that corresponds to the type of aircraft being flown and again in the column marked "Total." In this manner it is possible to record separately the pilot's experience with Fixed-Wing, Rotary and Lighter-Than-Air (LTA) UAV aircraft and to sum the total flight time.

Function Enter the duration of the flight, in hours and minutes or hours and tenths, in the column that corresponds to the role performed by the pilot during the flight. In this manner it is possible to record whether the flying was done as Operator in Charge (OIC) / Solo, or while receiving Training or acting as Instructor, separately from the total flight time.

Data Enter the duration of the portion of the flight that occurred at night.

Enter the number of landings performed during the flight.

Enter the distance flown. State your units.

Enter the maximum height achieved during the flight (or the height appropriate for the purpose of the flight). State your units.

Notes on the recording of pilot flight data (on even-numbered pages) in this logbook, continued:

Totals across the bottom of each page:

Totals this page: Sum the durations and landings entries from the columns above and enter these data in the spaces provided.

Totals brought forward: Enter the '**Totals to date**' data from the *previous page* in the spaces provided.

Totals to Date Sum the 'Totals this page' and the 'Totals brought forward' entries and enter these data into the spaces provided. Note that the resultant '**Totals to date**' for Fixed Wing plus Rotary plus LTA flight time should equal the entry for 'Total' time (bold-bordered box). The '**Totals to date**' for 'OIC/Solo' plus 'Training Received' and 'Instructor' flight time should equal the entry for 'Total' time as well.

Note also that the sums of Night flight time and the Landings accomplished have no effect on 'Total' time.

Sign the page in the lower-left corner in the space provided to attest to the fact that the entries are true and accurate.

Note that the grey color bands across each page are for keeping track of which line is being filled out. They have no other meaning.

An example is provided on the next page.

FLT. NO.	DATE Yr 2018 Dd / Mm	AIRCRAFT Make/Model Registration No.	FLIGHT LOCATION Takeoff Time / Landing Time			FLIGHT DURATION Fixed Wing	Rotary	LTA	Total	FUNCTION OIC/Solo	Training Received	Instructor	Night	DATA Ldgs.	Dist. m	Max. Ht. ft
9	01_05	DJI P3P N12345	Family farm, Anywhere	0800 L	0830 L		0:30		0:30	0:30				1	600	100
10	02_05	My Own N54321	364609N 104131?W	2210	2245	0:35			0:35			0:35	0:35	1	550	80

Example page:

In the example above the first two entries of a page and separately, below them, the corresponding totals, are shown as if the entire page had been filled-out and eight flights flown and entered.

The pilot has flown both Fixed-wing and Rotary aircraft and has logged the times separately. The duration of each flight has also been entered into the 'Total' column, so that the combined time of all aircraft flown can be calculated.

This pilot has entered the 'Totals this page' in the spaces provided at the bottom and has copied the corresponding '**Totals to date**' from the previous page and entered them here as 'Totals brought forward.'

The Fixed-wing time flown to date is therefore shown to be 7 hours 45 minutes, the total Rotary time to date is 5 hours 55 minutes and the pilot's absolute total time is 13 hours 40 minutes for all aircraft together.

Note that this pilot has flown 1 hour 55 minutes while giving instruction and 11 hours 45 minutes as OIC / Solo, which also sum to 13 hours 40 minutes. Across the top of the page the pilot has annotated "L" (local) as the time zone, "m" (meters) as the distance unit and "ft" (feet) as the maximum height unit.

			Totals this page:	4:15	3:40		7:55	6:00		1:55	0:35	8
			Totals brought forward:	3:30	2:15		5:45	5:45		0	1:20	8
I certify that the entries in this log are true. *J. Smith*			**Totals to date:**	7:45	5:55		13:40	11:45		1:55	1:55	16

NOTES CONCERNING THE FLIGHTS LOGGED ON THE PAGE ABOVE

Flt. No.	Batteries / Fuel	P-f. Ins. / C. Cal.	Technical Problems Found / Corrective Action, Status	Remarks and Endorsements
9	*Batt. A volts:*	Y	*Aircraft would not fly a straight line.*	*Satisfactory video data obtained after performing calibrations.*
	17.1 – 14.6	Y	*Made IMU and controller stick calibrations.*	*Client was satisfied and has asked for additional data.*

Example and Notes on the recording of technical data and remarks (on odd-numbered pages) in this logbook:

Flt. No.   Enter the corresponding flight number from the previous page above. The flight number entered in this example, therefore, corresponds to a flight recorded on the previous example page: flight number 9.

Batteries / Fuel   Enter amount of fuel uplifted and burned (for liquid-fueled aircraft) or battery charge and discharge values (in volts or % charged). Be sure to state the battery designation (e.g. name, number) recorded in the Power Source Data pages. In this example the pilot recorded that battery "A" had a charge of 17.1 volts prior to flight and a post-flight charge of 14.6 volts remaining.

P-f. I. / C. Cal.   Pre-flight Inspection / Compass Calibration: remark, using a crew-member's initials or a "Y" for Yes, that a pre-flight inspection was accomplished. Remark also (Y / N) whether a compass calibration was performed or not. In the example here both remarks are "Y."

Technical Problems Found / Corrective Action, Status
   Enter a brief description of any mechanical defect or incident that occurred during the numbered flights (or pre-flight). Enter also a description of any repair or fix accomplished to resolve the problems. Comment about any test flight requirement.

Remarks / Endorsements
   Enter pertinent observations about the flight (e.g. observer name, incidents occurrence, weather conditions affecting flight, etc), as well as instructor/examiner signature when appropriate, in this section.

FLT. NO.	DATE		AIRCRAFT		FLIGHT LOCATION			FLIGHT DURATION				FUNCTION			DATA			
	Yr ___	Dd / Mm	Make/Model	Registration No.	Takeoff Time	/	Landing Time	Fixed Wing	Rotary	LTA	Total	OIC/Solo	Training Received	Instructor	Night	Ldgs.	Dist.	Max. Ht.
	_/_																	
	_/_																	
	_/_																	
	_/_																	
	_/_																	
	_/_																	
	_/_																	
	_/_																	
	_/_																	

Totals this page:

Totals brought forward:

Totals to date:

I certify that the entries in this log are true.

_____

NOTES CONCERNING THE FLIGHTS LOGGED ON THE PAGE ABOVE

Flt. No.	Batteries / Fuel	P-f. Ins. / C. Cal.	Technical Problems Found / Corrective Action, Status	Remarks and Endorsements

FLT. NO.	DATE		AIRCRAFT		FLIGHT LOCATION		FLIGHT DURATION				FUNCTION			DATA			
	Yr ___		Make/Model		Takeoff / Landing		Fixed Wing	Rotary	LTA	Total	OIC/Solo	Training Received	Instructor	Night	Ldgs.	Dist.	Max. Ht.
	Dd / Mm		Registration No.		Time	Time											
	_/_																
	_/_																
	_/_																
	_/_																
	_/_																
	_/_																
	_/_																
	_/_																

I certify that the entries in this log are true.

Totals this page:

Totals brought forward:

Totals to date:

NOTES CONCERNING THE FLIGHTS LOGGED ON THE PAGE ABOVE

Flt. No.	Batteries / Fuel	P-f. Ins. — — C. Cal.	Technical Problems Found / Corrective Action, Status	Remarks and Endorsements

FLT. NO.	DATE		AIRCRAFT		FLIGHT LOCATION				FLIGHT DURATION				FUNCTION			DATA			
	Yr____		Make/Model		Takeoff	/	Landing		Fixed Wing	Rotary	LTA	Total	OIC/Solo	Training Received	Instructor	Night	Ldgs.	Dist.	Max. Ht.
	Dd / Mm		Registration No.		Time		Time												
	/																		
	/																		
	/																		
	/																		
	/																		
	/																		
	/																		
	/																		
	/																		
								Totals this page:											
			I certify that the entries in this log are true.					Totals brought forward:											
			_____					**Totals to date:**											

Flt. No.	Batteries / Fuel	P-f. Ins. — C. Cal.	NOTES CONCERNING THE FLIGHTS LOGGED ON THE PAGE ABOVE	
			Technical Problems Found / Corrective Action, Status	Remarks and Endorsements

FLT. NO.	DATE		AIRCRAFT		FLIGHT LOCATION		FLIGHT DURATION				FUNCTION			DATA			
	Yr ___	Dd / Mm	Make/Model	Registration No.	Takeoff Time	Landing Time	Fixed Wing	Rotary	LTA	Total	OIC/Solo	Training Received	Instructor	Night	Ldgs.	Dist.	Max. Ht.
	/				/												
	/				/												
	/				/												
	/				/												
	/				/												
	/				/												
	/				/												
	/				/												
	I certify that the entries in this log are true.					Totals this page:											
						Totals brought forward:											
						Totals to date:											

NOTES CONCERNING THE FLIGHTS LOGGED ON THE PAGE ABOVE

Flt. No.	Batteries / Fuel	P-f. Ins. / C. Cal.	Technical Problems Found / Corrective Action, Status	Remarks and Endorsements

FLT. NO.	DATE		AIRCRAFT		FLIGHT LOCATION			FLIGHT DURATION				FUNCTION			DATA			
	Yr	Dd/Mm	Make/Model	Registration No.	Takeoff Time	/	Landing Time	Fixed Wing	Rotary	LTA	Total	OIC/Solo	Training Received	Instructor	Night	Ldgs.	Dist.	Max. Ht.
		/				/												
		/				/												
		/				/												
		/				/												
		/				/												
		/				/												
		/				/												
		/				/												
	I certify that the entries in this log are true.						Totals this page:											
							Totals brought forward:											
							Totals to date:											

NOTES CONCERNING THE FLIGHTS LOGGED ON THE PAGE ABOVE

Flt. No.	Batteries / Fuel	P-f. Ins. / C. Cal.	Technical Problems Found / Corrective Action, Status	Remarks and Endorsements

FLT. NO.	DATE		AIRCRAFT		FLIGHT LOCATION		FLIGHT DURATION				FUNCTION			DATA			
	Yr___		Make/Model		Takeoff / Landing		Fixed Wing	Rotary	LTA	Total	OIC/Solo	Training Received	Instructor	Night	Ldgs.	Dist.	Max. Ht.
	Dd / Mm		Registration No.		Time	Time											
	_/_																
	_/_																
	_/_																
	_/_																
	_/_																
	_/_																
	_/_																
	_/_																
	I certify that the entries in this log are true.						Totals this page:										
							Totals brought forward:										
	_____						**Totals to date:**										

NOTES CONCERNING THE FLIGHTS LOGGED ON THE PAGE ABOVE

Flt. No.	Batteries / Fuel	P-f. Ins. – – – C. Cal.	Technical Problems Found / Corrective Action, Status	Remarks and Endorsements

FLT. NO.	DATE		AIRCRAFT		FLIGHT LOCATION			FLIGHT DURATION				FUNCTION			DATA			
	Yr ___		Make/Model		Takeoff	/	Landing	Fixed Wing	Rotary	LTA	Total	OIC/Solo	Training Received	Instructor	Night	Ldgs.	Dist.	Max. Ht.
	Dd / Mm		Registration No.		Time		Time											
	_/_																	
	_/_																	
	_/_																	
	_/_																	
	_/_																	
	_/_																	
	_/_																	
	_/_																	
	_/_																	
							Totals this page:											
	I certify that the entries in this log are true.						Totals brought forward:											
	_____						**Totals to date:**											

NOTES CONCERNING THE FLIGHTS LOGGED ON THE PAGE ABOVE

Flt. No.	Batteries / Fuel	P-f. Ins. / C. Cal.	Technical Problems Found / Corrective Action, Status	Remarks and Endorsements

FLT. NO.	DATE		AIRCRAFT		FLIGHT LOCATION		FLIGHT DURATION				FUNCTION			DATA			
	Yr ___		Make/Model		Takeoff	/ Landing	Fixed Wing	Rotary	LTA	Total	OIC/Solo	Training Received	Instructor	Night	Ldgs.	Dist.	Max. Ht.
	Dd / Mm		Registration No.		Time	Time											
	_/_					/											
	_/_					/											
	_/_					/											
	_/_					/											
	_/_					/											
	_/_					/											
	_/_					/											
	_/_					/											
	_/_					/											
	I certify that the entries in this log are true. _____									Totals this page:							
										Totals brought forward:							
										Totals to date:							

Flt. No.	Batteries / Fuel	P-f. Ins. ― ― C. Cal.	Technical Problems Found / Corrective Action, Status	Remarks and Endorsements

NOTES CONCERNING THE FLIGHTS LOGGED ON THE PAGE ABOVE

FLT. NO.	DATE		AIRCRAFT		FLIGHT LOCATION		FLIGHT DURATION				FUNCTION			DATA			
	Yr ____		Make/Model	Takeoff	/	Landing	Fixed Wing	Rotary	LTA	Total	OIC/Solo	Training Received	Instructor	Night	Ldgs.	Dist.	Max. Ht.
	Dd / Mm		Registration No.	Time		Time											
	/																
	/																
	/																
	/																
	/																
	/																
	/																
	/																
	/																

I certify that the entries in this log are true.

_____

Totals this page:

Totals brought forward:

Totals to date:

Flt. No.	Batteries / Fuel	P-f. Ins. — — — C. Cal.	Technical Problems Found / Corrective Action, Status	Remarks and Endorsements

NOTES CONCERNING THE FLIGHTS LOGGED ON THE PAGE ABOVE

FLT. NO.	DATE		AIRCRAFT		FLIGHT LOCATION			FLIGHT DURATION				FUNCTION			DATA			
	Yr ___		Make/Model		Takeoff	/	Landing	Fixed Wing	Rotary	LTA	Total	OIC/Solo	Training Received	Instructor	Night	Ldgs.	Dist.	Max. Ht.
	Dd / Mm		Registration No.		Time		Time											
	/																	
	/																	
	/																	
	/																	
	/																	
	/																	
	/																	
	/																	
	/																	

I certify that the entries in this log are true.

_____

Totals this page:
Totals brought forward:
Totals to date:

Flt. No.	Batteries / Fuel	P-f. Ins. / C. Cal.	NOTES CONCERNING THE FLIGHTS LOGGED ON THE PAGE ABOVE	
			Technical Problems Found / Corrective Action, Status	Remarks and Endorsements

FLT. NO.	DATE		AIRCRAFT		FLIGHT LOCATION			FLIGHT DURATION				FUNCTION			DATA			
	Yr ___		Make/Model		Takeoff	/	Landing	Fixed Wing	Rotary	LTA	Total	OIC/Solo	Training Received	Instructor	Night	Ldgs.	Dist.	Max. Ht.
	Dd / Mm		Registration No.		Time		Time											
	_/_																	
	_/_																	
	_/_																	
	_/_																	
	_/_																	
	_/_																	
	_/_																	
	_/_																	
	_/_																	

Totals this page:
Totals brought forward:
Totals to date:

I certify that the entries in this log are true.

46

NOTES CONCERNING THE FLIGHTS LOGGED ON THE PAGE ABOVE

Flt. No.	Batteries / Fuel	P-f. Ins. / C. Cal.	Technical Problems Found / Corrective Action, Status	Remarks and Endorsements

FLT. NO.	DATE		AIRCRAFT	FLIGHT LOCATION			FLIGHT DURATION				FUNCTION			DATA			
	Yr ___		Make/Model	Takeoff	/	Landing	Fixed Wing	Rotary	LTA	Total	OIC/Solo	Training Received	Instructor	Night	Ldgs.	Dist.	Max. Ht.
	Dd / Mm		Registration No.	Time		Time											
	_/_																
	_/_																
	_/_																
	_/_																
	_/_																
	_/_																
	_/_																
	_/_																
	I certify that the entries in this log are true.									Totals this page:							
										Totals brought forward:							
	_____									**Totals to date:**							

NOTES CONCERNING THE FLIGHTS LOGGED ON THE PAGE ABOVE

Flt. No.	Batteries / Fuel	P-f. Ins. / C. Cal.	Technical Problems Found / Corrective Action, Status	Remarks and Endorsements

FLT. NO.	DATE		AIRCRAFT		FLIGHT LOCATION			FLIGHT DURATION				FUNCTION			DATA			
	Yr___		Make/Model		Takeoff	/	Landing	Fixed Wing	Rotary	LTA	Total	OIC/Solo	Training Received	Instructor	Night	Ldgs.	Dist.	Max. Ht.
	Dd / Mm		Registration No.		Time		Time											
	_/_																	
	_/_																	
	_/_																	
	_/_																	
	_/_																	
	_/_																	
	_/_																	
	_/_																	
	_/_																	
							Totals this page:											
							Totals brought forward:											
	I certify that the entries in this log are true. _____						**Totals to date:**											

50

NOTES CONCERNING THE FLIGHTS LOGGED ON THE PAGE ABOVE

Flt. No.	Batteries / Fuel	P-f. Ins. / C. Cal.	Technical Problems Found / Corrective Action, Status	Remarks and Endorsements

FLT. NO.	DATE		AIRCRAFT		FLIGHT LOCATION		FLIGHT DURATION				FUNCTION			DATA			
	Yr____	Dd / Mm	Make/Model	Registration No.	Takeoff Time	Landing Time	Fixed Wing	Rotary	LTA	Total	OIC/Solo	Training Received	Instructor	Night	Ldgs.	Dist.	Max. Ht.
	/				/												
	/				/												
	/				/												
	/				/												
	/				/												
	/				/												
	/				/												
	/				/												

I certify that the entries in this log are true.

_____

Totals this page:
Totals brought forward:
Totals to date:

NOTES CONCERNING THE FLIGHTS LOGGED ON THE PAGE ABOVE

Flt. No.	Batteries / Fuel	P-f. Ins. / C. Cal.	Technical Problems Found / Corrective Action, Status	Remarks and Endorsements

FLT. NO.	DATE		AIRCRAFT		FLIGHT LOCATION			FLIGHT DURATION				FUNCTION			DATA			
	Yr ___		Make/Model		Takeoff /		Landing	Fixed Wing	Rotary	LTA	Total	OIC/Solo	Training Received	Instructor	Night	Ldgs.	Dist.	Max. Ht.
	Dd / Mm		Registration No.		Time		Time											
	/				/													
	/				/													
	/				/													
	/				/													
	/				/													
	/				/													
	/				/													
	/				/													
	/				/													

Totals this page:

Totals brought forward:

Totals to date:

I certify that the entries in this log are true.

_____

54

NOTES CONCERNING THE FLIGHTS LOGGED ON THE PAGE ABOVE

Flt. No.	Batteries / Fuel	P-f. Ins. / C. Cal.	Technical Problems Found / Corrective Action, Status	Remarks and Endorsements

FLT. NO.	DATE		AIRCRAFT		FLIGHT LOCATION			FLIGHT DURATION				FUNCTION			DATA			
	Yr ___	Dd / Mm	Make/Model	Registration No.	Takeoff Time	/	Landing Time	Fixed Wing	Rotary	LTA	Total	OIC/Solo	Training Received	Instructor	Night	Ldgs.	Dist.	Max. Ht.
	___/___																	
	___/___																	
	___/___																	
	___/___																	
	___/___																	
	___/___																	
	___/___																	
	___/___																	

I certify that the entries in this log are true.

Totals this page:

Totals brought forward:

Totals to date:

NOTES CONCERNING THE FLIGHTS LOGGED ON THE PAGE ABOVE

Flt. No.	Batteries / Fuel	P-f. Ins. / C. Cal.	Technical Problems Found / Corrective Action, Status	Remarks and Endorsements

FLT. NO.	DATE		AIRCRAFT		FLIGHT LOCATION			FLIGHT DURATION				FUNCTION			DATA			
	Yr___	Dd / Mm	Make/Model	Registration No.	Takeoff Time	/	Landing Time	Fixed Wing	Rotary	LTA	Total	OIC/Solo	Training Received	Instructor	Night	Ldgs.	Dist.	Max. Ht.
	_/_					/												
	_/_					/												
	_/_					/												
	_/_					/												
	_/_					/												
	_/_					/												
	_/_					/												
	_/_					/												

Totals this page:
Totals brought forward:
Totals to date:

I certify that the entries in this log are true.

_____

			NOTES CONCERNING THE FLIGHTS LOGGED ON THE PAGE ABOVE	
Flt. No.	Batteries / Fuel	P-f. Ins. — — C. Cal.	Technical Problems Found / Corrective Action, Status	Remarks and Endorsements

FLT. NO.	DATE		AIRCRAFT		FLIGHT LOCATION			FLIGHT DURATION				FUNCTION			DATA			
	Yr ___	Dd / Mm	Make/Model	Registration No.	Takeoff Time	/	Landing Time	Fixed Wing	Rotary	LTA	Total	OIC/Solo	Training Received	Instructor	Night	Ldgs.	Dist.	Max. Ht.
	_/_																	
	_/_																	
	_/_																	
	_/_																	
	_/_																	
	_/_																	
	_/_																	
	_/_																	

I certify that the entries in this log are true.

Totals this page:

Totals brought forward:

Totals to date:

Flt. No.	Batteries / Fuel	P-f. Ins. – – – C. Cal.	Technical Problems Found / Corrective Action, Status	Remarks and Endorsements

NOTES CONCERNING THE FLIGHTS LOGGED ON THE PAGE ABOVE

FLT. NO.	DATE		AIRCRAFT		FLIGHT LOCATION			FLIGHT DURATION				FUNCTION			DATA			
	Yr___		Make/Model		Takeoff	/	Landing	Fixed Wing	Rotary	LTA	Total	OIC/Solo	Training Received	Instructor	Night	Ldgs.	Dist.	Max. Ht.
	Dd / Mm		Registration No.		Time		Time											
	/																	
	/																	
	/																	
	/																	
	/																	
	/																	
	/																	
	/																	
	/																	

I certify that the entries in this log are true.

_____

Totals this page:
Totals brought forward:
Totals to date:

NOTES CONCERNING THE FLIGHTS LOGGED ON THE PAGE ABOVE

Flt. No.	Batteries / Fuel	P-f. Ins. / C. Cal.	Technical Problems Found / Corrective Action, Status	Remarks and Endorsements

FLT. NO.	DATE		AIRCRAFT		FLIGHT LOCATION			FLIGHT DURATION				FUNCTION			DATA			
	Yr ____		Make/Model		Takeoff	/	Landing	Fixed Wing	Rotary	LTA	Total	OIC/Solo	Training Received	Instructor	Night	Ldgs.	Dist.	Max. Ht.
	Dd / Mm		Registration No.		Time		Time											
	_/_																	
	_/_																	
	_/_																	
	_/_																	
	_/_																	
	_/_																	
	_/_																	
	_/_																	
							Totals this page:											
							Totals brought forward:											
	I certify that the entries in this log are true.						**Totals to date:**											

Flt. No.	Batteries / Fuel	P-f. Ins. — — C. Cal.	NOTES CONCERNING THE FLIGHTS LOGGED ON THE PAGE ABOVE	
			Technical Problems Found / Corrective Action, Status	Remarks and Endorsements

FLT. NO.	DATE		AIRCRAFT		FLIGHT LOCATION			FLIGHT DURATION				FUNCTION			DATA			
	Yr ___	Dd / Mm	Make/Model	Registration No.	Takeoff Time	/	Landing Time	Fixed Wing	Rotary	LTA	Total	OIC/Solo	Training Received	Instructor	Night	Ldgs.	Dist.	Max. Ht.
		_/_																
		_/_																
		_/_																
		_/_																
		_/_																
		_/_																
		_/_																
		_/_																
		_/_																

Totals this page:
Totals brought forward:
Totals to date:

I certify that the entries in this log are true.

_____

NOTES CONCERNING THE FLIGHTS LOGGED ON THE PAGE ABOVE

Flt. No.	Batteries / Fuel	P-f. Ins. / C. Cal.	Technical Problems Found / Corrective Action, Status	Remarks and Endorsements

FLT. NO.	DATE		AIRCRAFT		FLIGHT LOCATION			FLIGHT DURATION				FUNCTION			DATA			
	Yr ___	Dd / Mm	Make/Model	Registration No.	Takeoff Time	/	Landing Time	Fixed Wing	Rotary	LTA	Total	OIC/Solo	Training Received	Instructor	Night	Ldgs.	Dist.	Max. Ht.
	___/___																	
	___/___																	
	___/___																	
	___/___																	
	___/___																	
	___/___																	
	___/___																	
	___/___																	
					Totals this page:													
					Totals brought forward:													
					Totals to date:													

I certify that the entries in this log are true.

NOTES CONCERNING THE FLIGHTS LOGGED ON THE PAGE ABOVE

Flt. No.	Batteries / Fuel	P-f. Ins. / C. Cal.	Technical Problems Found / Corrective Action, Status	Remarks and Endorsements

FLT. NO.	DATE		AIRCRAFT		FLIGHT LOCATION				FLIGHT DURATION				FUNCTION			DATA			
	Yr___		Make/Model		Takeoff	/	Landing		Fixed Wing	Rotary	LTA	Total	OIC/Solo	Training Received	Instructor	Night	Ldgs.	Dist.	Max. Ht.
	Dd / Mm		Registration No.		Time		Time												
	___/___																		
	___/___																		
	___/___																		
	___/___																		
	___/___																		
	___/___																		
	___/___																		
	___/___																		
	___/___																		

I certify that the entries in this log are true.

_____

Totals this page:

Totals brought forward:

Totals to date:

NOTES CONCERNING THE FLIGHTS LOGGED ON THE PAGE ABOVE

Flt. No.	Batteries / Fuel	P-f. Ins. / C. Cal.	Technical Problems Found / Corrective Action, Status	Remarks and Endorsements

FLT. NO.	DATE		AIRCRAFT		FLIGHT LOCATION		FLIGHT DURATION				FUNCTION			DATA			
	Yr ___		Make/Model		Takeoff /	Landing	Fixed Wing	Rotary	LTA	Total	OIC/Solo	Training Received	Instructor	Night	Ldgs.	Dist.	Max. Ht.
	Dd / Mm		Registration No.		Time	Time											
	_/_																
	_/_																
	_/_																
	_/_																
	_/_																
	_/_																
	_/_																
	_/_																
											Totals this page:						
I certify that the entries in this log are true.											Totals brought forward:						
_____											**Totals to date:**						

Flt. No.	Batteries / Fuel	P-f. Ins. — — C. Cal.	Technical Problems Found / Corrective Action, Status	Remarks and Endorsements

NOTES CONCERNING THE FLIGHTS LOGGED ON THE PAGE ABOVE

FLT. NO.	DATE		AIRCRAFT		FLIGHT LOCATION			FLIGHT DURATION				FUNCTION			DATA			
	Yr ___		Make/Model		Takeoff	/	Landing	Fixed Wing	Rotary	LTA	Total	OIC/Solo	Training Received	Instructor	Night	Ldgs.	Dist.	Max. Ht.
	Dd / Mm		Registration No.		Time		Time											
	/																	
	/																	
	/																	
	/																	
	/																	
	/																	
	/																	
	/																	
	/																	
	I certify that the entries in this log are true. _____							Totals this page:										
								Totals brought forward:										
								Totals to date:										

NOTES CONCERNING THE FLIGHTS LOGGED ON THE PAGE ABOVE

Flt. No.	Batteries / Fuel	P-f. Ins. − − − C. Cal.	Technical Problems Found / Corrective Action, Status	Remarks and Endorsements

FLT. NO.	DATE Yr ___ Dd / Mm	AIRCRAFT Make/Model / Registration No.	FLIGHT LOCATION Takeoff Time	FLIGHT LOCATION Landing Time	FLIGHT DURATION Fixed Wing	FLIGHT DURATION Rotary	FLIGHT DURATION LTA	FLIGHT DURATION Total	FUNCTION OIC/Solo	FUNCTION Training Received	FUNCTION Instructor	DATA Night	DATA Ldgs.	DATA Dist.	DATA Max. Ht.
	___/___														
	___/___														
	___/___														
	___/___														
	___/___														
	___/___														
	___/___														
	___/___														

I certify that the entries in this log are true.

_____

Totals this page:

Totals brought forward:

Totals to date:

NOTES CONCERNING THE FLIGHTS LOGGED ON THE PAGE ABOVE

Flt. No.	Batteries / Fuel	P-f. Ins. / C. Cal.	Technical Problems Found / Corrective Action, Status	Remarks and Endorsements

FLT. NO.	DATE		AIRCRAFT		FLIGHT LOCATION			FLIGHT DURATION				FUNCTION			DATA			
	Yr___		Make/Model		Takeoff	/	Landing	Fixed Wing	Rotary	LTA	Total	OIC/Solo	Training Received	Instructor	Night	Ldgs.	Dist.	Max. Ht.
	Dd / Mm		Registration No.		Time		Time											
	_/_																	
	_/_																	
	_/_																	
	_/_																	
	_/_																	
	_/_																	
	_/_																	
	_/_																	
	_/_																	
							Totals this page:											
							Totals brought forward:											
							Totals to date:											

I certify that the entries in this log are true.

_____

		NOTES CONCERNING THE FLIGHTS LOGGED ON THE PAGE ABOVE		
Flt. No.	Batteries / Fuel	P-f. Ins. − − − C. Cal.	Technical Problems Found / Corrective Action, Status	Remarks and Endorsements

FLT. NO.	DATE		AIRCRAFT	FLIGHT LOCATION			FLIGHT DURATION				FUNCTION			DATA			
	Yr ___		Make/Model	Takeoff	/	Landing	Fixed Wing	Rotary	LTA	Total	OIC/Solo	Training Received	Instructor	Night	Ldgs.	Dist.	Max. Ht.
	Dd / Mm		Registration No.	Time		Time											
	/																
	/																
	/																
	/																
	/																
	/																
	/																
	/																
	I certify that the entries in this log are true.						Totals this page:										
							Totals brought forward:										
							Totals to date:										

Flt. No.	Batteries / Fuel	P-f. Ins. / C. Cal.	Technical Problems Found / Corrective Action, Status	Remarks and Endorsements

NOTES CONCERNING THE FLIGHTS LOGGED ON THE PAGE ABOVE

FLT. NO.	DATE		AIRCRAFT		FLIGHT LOCATION				FLIGHT DURATION				FUNCTION			DATA			
	Yr ___	Dd / Mm	Make/Model	Registration No.	Takeoff	Time	/	Landing	Time	Fixed Wing	Rotary	LTA	Total	OIC/Solo	Training Received	Instructor	Night	Ldgs.	Dist.	Max. Ht.
	/					/												
	/					/												
	/					/												
	/					/												
	/					/												
	/					/												
	/					/												
	/					/												
								Totals this page:										
								Totals brought forward:										
	I certify that the entries in this log are true.							**Totals to date:**										

NOTES CONCERNING THE FLIGHTS LOGGED ON THE PAGE ABOVE

Flt. No.	Batteries / Fuel	P-f. Ins. — — C. Cal.	Technical Problems Found / Corrective Action, Status	Remarks and Endorsements

FLT. NO.	DATE		AIRCRAFT		FLIGHT LOCATION		FLIGHT DURATION				FUNCTION			DATA			
	Yr ___		Make/Model		Takeoff / Landing		Fixed Wing	Rotary	LTA	Total	OIC/Solo	Training Received	Instructor	Night	Ldgs.	Dist.	Max. Ht.
	Dd / Mm		Registration No.		Time	Time											
	_/_																
	_/_																
	_/_																
	_/_																
	_/_																
	_/_																
	_/_																
	_/_																

I certify that the entries in this log are true.

_____

Totals this page:

Totals brought forward:

Totals to date:

NOTES CONCERNING THE FLIGHTS LOGGED ON THE PAGE ABOVE

Flt. No.	Batteries / Fuel	P-f. Ins. / C. Cal.	Technical Problems Found / Corrective Action, Status	Remarks and Endorsements

FLT. NO.	DATE		AIRCRAFT		FLIGHT LOCATION			FLIGHT DURATION				FUNCTION			DATA			
	Yr ___		Make/Model		Takeoff	/	Landing	Fixed Wing	Rotary	LTA	Total	OIC/Solo	Training Received	Instructor	Night	Ldgs.	Dist.	Max. Ht.
	Dd / Mm		Registration No.		Time		Time											
	_/_																	
	_/_																	
	_/_																	
	_/_																	
	_/_																	
	_/_																	
	_/_																	
	_/_																	
	_/_																	
							Totals this page:											
			I certify that the entries in this log are true.				Totals brought forward:											
			_____				**Totals to date:**											

Flt. No.	Batteries / Fuel	P-f. Ins. / C. Cal.	NOTES CONCERNING THE FLIGHTS LOGGED ON THE PAGE ABOVE	
			Technical Problems Found / Corrective Action, Status	Remarks and Endorsements

FLT. NO.	DATE		AIRCRAFT		FLIGHT LOCATION			FLIGHT DURATION				FUNCTION			DATA			
	Yr		Make/Model		Takeoff	/	Landing	Fixed Wing	Rotary	LTA	Total	OIC/Solo	Training Received	Instructor	Night	Ldgs.	Dist.	Max. Ht.
	Dd / Mm		Registration No.		Time		Time											
	_/_																	
	_/_																	
	_/_																	
	_/_																	
	_/_																	
	_/_																	
	_/_																	
	_/_																	
							Totals this page:											
			I certify that the entries in this log are true.				Totals brought forward:											
			_____				**Totals to date:**											

Flt. No.	Batteries / Fuel	P-f. Ins. / C. Cal.	NOTES CONCERNING THE FLIGHTS LOGGED ON THE PAGE ABOVE	
			Technical Problems Found / Corrective Action, Status	Remarks and Endorsements

FLT. NO.	DATE		AIRCRAFT		FLIGHT LOCATION		FLIGHT DURATION				FUNCTION			DATA				
	Yr___		Make/Model		Takeoff	/ Landing	Fixed Wing	Rotary	LTA	Total	OIC/Solo	Training Received	Instructor	Night	Ldgs.	Dist.	Max. Ht.	
	Dd / Mm		Registration No.		Time	Time												
	_/_																	
	_/_																	
	_/_																	
	_/_																	
	_/_																	
	_/_																	
	_/_																	
	_/_																	
	_/_																	

Totals this page:

Totals brought forward:

Totals to date:

I certify that the entries in this log are true.

_____

NOTES CONCERNING THE FLIGHTS LOGGED ON THE PAGE ABOVE

Flt. No.	Batteries / Fuel	P-f. Ins. / C. Cal.	Technical Problems Found / Corrective Action, Status	Remarks and Endorsements

FLT. NO.	DATE		AIRCRAFT		FLIGHT LOCATION		FLIGHT DURATION				FUNCTION			DATA			
	Yr ___		Make/Model		Takeoff	/ Landing	Fixed Wing	Rotary	LTA	Total	OIC/Solo	Training Received	Instructor	Night	Ldgs.	Dist.	Max. Ht.
	Dd / Mm		Registration No.		Time	Time											
	_/_																
	_/_																
	_/_																
	_/_																
	_/_																
	_/_																
	_/_																
	_/_																

I certify that the entries in this log are true.

Totals this page:

Totals brought forward:

Totals to date:

_____

Flt. No.	Batteries / Fuel	P-f. Ins. — — C. Cal.	Technical Problems Found / Corrective Action, Status	Remarks and Endorsements

NOTES CONCERNING THE FLIGHTS LOGGED ON THE PAGE ABOVE

FLT. NO.	DATE		AIRCRAFT		FLIGHT LOCATION			FLIGHT DURATION				FUNCTION			DATA			
	Yr ___	Dd / Mm	Make/Model	Registration No.	Takeoff Time	/	Landing Time	Fixed Wing	Rotary	LTA	Total	OIC/Solo	Training Received	Instructor	Night	Ldgs.	Dist.	Max. Ht.
	/																	
	/																	
	/																	
	/																	
	/																	
	/																	
	/																	
	/																	
	/																	

I certify that the entries in this log are true.

_____

Totals this page:
Totals brought forward:
Totals to date:

Flt. No.	Batteries / Fuel	P-f. Ins. — C. Cal.	Technical Problems Found / Corrective Action, Status	Remarks and Endorsements
			NOTES CONCERNING THE FLIGHTS LOGGED ON THE PAGE ABOVE	

FLT. NO.	DATE		AIRCRAFT		FLIGHT LOCATION				FLIGHT DURATION				FUNCTION			DATA			
	Yr ___	Dd / Mm	Make/Model	Registration No.	Takeoff Time	/	Landing Time		Fixed Wing	Rotary	LTA	Total	OIC/Solo	Training Received	Instructor	Night	Ldgs.	Dist.	Max. Ht.
	_/_																		
	_/_																		
	_/_																		
	_/_																		
	_/_																		
	_/_																		
	_/_																		
	_/_																		
	_/_																		

I certify that the entries in this log are true.

Totals this page:

Totals brought forward:

Totals to date:

NOTES CONCERNING THE FLIGHTS LOGGED ON THE PAGE ABOVE

Flt. No.	Batteries / Fuel	P-f. Ins. / C. Cal.	Technical Problems Found / Corrective Action, Status	Remarks and Endorsements

FLT. NO.	DATE		AIRCRAFT		FLIGHT LOCATION			FLIGHT DURATION				FUNCTION			DATA			
	Yr___		Make/Model		Takeoff	/	Landing	Fixed Wing	Rotary	LTA	Total	OIC/Solo	Training Received	Instructor	Night	Ldgs.	Dist.	Max. Ht.
	Dd / Mm		Registration No.		Time		Time											
	_/_																	
	_/_																	
	_/_																	
	_/_																	
	_/_																	
	_/_																	
	_/_																	
	_/_																	

Totals this page:
Totals brought forward:
Totals to date:

I certify that the entries in this log are true.

Flt. No.	Batteries / Fuel	P-f. Ins. — C. Cal.	NOTES CONCERNING THE FLIGHTS LOGGED ON THE PAGE ABOVE	
			Technical Problems Found / Corrective Action, Status	Remarks and Endorsements

FLT. No.	DATE		AIRCRAFT		FLIGHT LOCATION		FLIGHT DURATION				FUNCTION			DATA			
	Yr ___		Make/Model		Takeoff /	Landing	Fixed Wing	Rotary	LTA	Total	OIC/Solo	Training Received	Instructor	Night	Ldgs.	Dist.	Max. Ht.
	Dd / Mm		Registration No.		Time	Time											
	_/_																
	_/_																
	_/_																
	_/_																
	_/_																
	_/_																
	_/_																
	_/_																
	_/_																
							Totals this page:										
	I certify that the entries in this log are true.						Totals brought forward:										
	_____						**Totals to date:**										

Flt. No.	Batteries / Fuel	P-f. Ins. — — C. Cal.	NOTES CONCERNING THE FLIGHTS LOGGED ON THE PAGE ABOVE	
			Technical Problems Found / Corrective Action, Status	Remarks and Endorsements

FLT. NO.	DATE		AIRCRAFT		FLIGHT LOCATION			FLIGHT DURATION				FUNCTION			DATA			
	Yr ___	Dd / Mm	Make/Model	Registration No.	Takeoff Time	/	Landing Time	Fixed Wing	Rotary	LTA	Total	OIC/Solo	Training Received	Instructor	Night	Ldgs.	Dist.	Max. Ht.
		___/___				/												
		___/___				/												
		___/___				/												
		___/___				/												
		___/___				/												
		___/___				/												
		___/___				/												
		___/___				/												
		___/___				/												
	I certify that the entries in this log are true. _____										**Totals this page:**							
											Totals brought forward:							
											Totals to date:							

NOTES CONCERNING THE FLIGHTS LOGGED ON THE PAGE ABOVE

Flt. No.	Batteries / Fuel	P.-f. Ins. / C. Cal.	Technical Problems Found / Corrective Action, Status	Remarks and Endorsements

FLT. NO.	DATE		AIRCRAFT		FLIGHT LOCATION			FLIGHT DURATION				FUNCTION			DATA			
	Yr ____	Dd / Mm	Make/Model	Registration No.	Takeoff Time	/	Landing Time	Fixed Wing	Rotary	LTA	Total	OIC/Solo	Training Received	Instructor	Night	Ldgs.	Dist.	Max. Ht.
	_ / _																	
	_ / _																	
	_ / _																	
	_ / _																	
	_ / _																	
	_ / _																	
	_ / _																	
	_ / _																	
	_ / _																	

I certify that the entries in this log are true.

_____

Totals this page:

Totals brought forward:

Totals to date:

NOTES CONCERNING THE FLIGHTS LOGGED ON THE PAGE ABOVE

Flt. No.	Batteries / Fuel	P-f. Ins. / C. Cal.	Technical Problems Found / Corrective Action, Status	Remarks and Endorsements

FLT. NO.	DATE		AIRCRAFT		FLIGHT LOCATION			FLIGHT DURATION				FUNCTION			DATA			
	Yr___	Dd / Mm	Make/Model	Registration No.	Takeoff Time	/	Landing Time	Fixed Wing	Rotary	LTA	Total	OIC/Solo	Training Received	Instructor	Night	Ldgs.	Dist.	Max. Ht.
	_/_																	
	_/_																	
	_/_																	
	_/_																	
	_/_																	
	_/_																	
	_/_																	
	_/_																	
	_/_																	

Totals this page:
Totals brought forward:
Totals to date:

I certify that the entries in this log are true.

_____

Flt. No.	Batteries / Fuel	P.-f. Ins. – – – C. Cal.	Technical Problems Found / Corrective Action, Status	Remarks and Endorsements

NOTES CONCERNING THE FLIGHTS LOGGED ON THE PAGE ABOVE

FLT. No.	DATE		AIRCRAFT		FLIGHT LOCATION			FLIGHT DURATION				FUNCTION			DATA			
	Yr ___	Dd / Mm	Make/Model	Registration No.	Takeoff Time	/ Landing Time		Fixed Wing	Rotary	LTA	Total	OIC/Solo	Training Received	Instructor	Night	Ldgs.	Dist.	Max. Ht.
	_/_																	
	_/_																	
	_/_																	
	_/_																	
	_/_																	
	_/_																	
	_/_																	
	_/_																	
	_/_																	
	I certify that the entries in this log are true. _____				Totals this page:													
					Totals brought forward:													
					Totals to date:													

NOTES CONCERNING THE FLIGHTS LOGGED ON THE PAGE ABOVE

Flt. No.	Batteries / Fuel	P-f. Ins. / C. Cal.	Technical Problems Found / Corrective Action, Status	Remarks and Endorsements

FLT. NO.	DATE		AIRCRAFT		FLIGHT LOCATION			FLIGHT DURATION				FUNCTION			DATA			
	Yr___	Dd / Mm	Make/Model	Registration No.	Takeoff Time	/	Landing Time	Fixed Wing	Rotary	LTA	Total	OIC/Solo	Training Received	Instructor	Night	Ldgs.	Dist.	Max. Ht.
	___/___					/												
	___/___					/												
	___/___					/												
	___/___					/												
	___/___					/												
	___/___					/												
	___/___					/												
	___/___					/												
	___/___					/												

Totals this page:
Totals brought forward:
Totals to date:

I certify that the entries in this log are true.

Notes concerning the flights logged on the page above

Flt. No.	Batteries / Fuel	P-f. Ins. / C. Cal.	Technical Problems Found / Corrective Action, Status	Remarks and Endorsements

FLT. No.	DATE Yr ___ Dd / Mm	AIRCRAFT Make/Model / Registration No.	FLIGHT LOCATION Takeoff Time	Landing Time	FLIGHT DURATION Fixed Wing	Rotary	LTA	Total	FUNCTION OIC/Solo	Training Received	Instructor	DATA Night	Ldgs.	Dist.	Max. Ht.
	/														
	/														
	/														
	/														
	/														
	/														
	/														
	/														
			Totals this page:												
			Totals brought forward:												
			Totals to date:												

I certify that the entries in this log are true.

_____

NOTES CONCERNING THE FLIGHTS LOGGED ON THE PAGE ABOVE

Flt. No.	Batteries / Fuel	P-f. Ins. / C. Cal.	Technical Problems Found / Corrective Action, Status	Remarks and Endorsements

FLT. NO.	DATE		AIRCRAFT		FLIGHT LOCATION				FLIGHT DURATION				FUNCTION			DATA			
	Yr ___	Dd / Mm	Make/Model	Registration No.	Takeoff	/	Landing		Fixed Wing	Rotary	LTA	Total	OIC/Solo	Training Received	Instructor	Night	Ldgs.	Dist.	Max. Ht.
					Time		Time												
	___/___																		
	___/___																		
	___/___																		
	___/___																		
	___/___																		
	___/___																		
	___/___																		
	___/___																		
	___/___																		
	I certify that the entries in this log are true.							Totals this page:											
								Totals brought forward:											
	_____							**Totals to date:**											

NOTES CONCERNING THE FLIGHTS LOGGED ON THE PAGE ABOVE

Flt. No.	Batteries / Fuel	P-f. Ins. / C. Cal.	Technical Problems Found / Corrective Action, Status	Remarks and Endorsements

FLT. NO.	DATE		AIRCRAFT		FLIGHT LOCATION			FLIGHT DURATION				FUNCTION			DATA			
	Yr		Make/Model		Takeoff	/	Landing	Fixed Wing	Rotary	LTA	Total	OIC/Solo	Training Received	Instructor	Night	Ldgs.	Dist.	Max. Ht.
	Dd / Mm		Registration No.		Time		Time											
	_/_																	
	_/_																	
	_/_																	
	_/_																	
	_/_																	
	_/_																	
	_/_																	
	_/_																	

I certify that the entries in this log are true.

Totals this page:

Totals brought forward:

Totals to date:

NOTES CONCERNING THE FLIGHTS LOGGED ON THE PAGE ABOVE

Flt. No.	Batteries / Fuel	P-f. Ins. — — C. Cal.	Technical Problems Found / Corrective Action, Status	Remarks and Endorsements

FLT. No.	DATE		AIRCRAFT		FLIGHT LOCATION			FLIGHT DURATION				FUNCTION			DATA			
	Yr____		Make/Model		Takeoff	/	Landing	Fixed Wing	Rotary	LTA	Total	OIC/Solo	Training Received	Instructor	Night	Ldgs.	Dist.	Max. Ht.
	Dd / Mm		Registration No.		Time		Time											
	/																	
	/																	
	/																	
	/																	
	/																	
	/																	
	/																	
	/																	
	/																	

I certify that the entries in this log are true.

_____

Totals this page:
Totals brought forward:
Totals to date:

NOTES CONCERNING THE FLIGHTS LOGGED ON THE PAGE ABOVE

Flt. No.	Batteries / Fuel	P-f. Ins. / C. Cal.	Technical Problems Found / Corrective Action, Status	Remarks and Endorsements

FLT. NO.	DATE Yr___ Dd / Mm	AIRCRAFT Make/Model / Registration No.	FLIGHT LOCATION Takeoff Time	/	Landing Time	FLIGHT DURATION Fixed Wing	Rotary	LTA	Total	FUNCTION OIC/Solo	Training Received	Instructor	DATA Night	Ldgs.	Dist.	Max. Ht.
	_/_															
	_/_															
	_/_															
	_/_															
	_/_															
	_/_															
	_/_															
	_/_															
	I certify that the entries in this log are true. _____					Totals this page:										
						Totals brought forward:										
						Totals to date:										

Flt. No.	Batteries / Fuel	P-f. Ins. / C. Cal.	NOTES CONCERNING THE FLIGHTS LOGGED ON THE PAGE ABOVE	
			Technical Problems Found / Corrective Action, Status	Remarks and Endorsements

FLT. NO.	DATE		AIRCRAFT		FLIGHT LOCATION				FLIGHT DURATION				FUNCTION			DATA				
	Yr ___		Make/Model		Takeoff		/	Landing	Fixed Wing	Rotary	LTA	Total	OIC/Solo	Training Received	Instructor	Night	Ldgs.	Dist.	Max. Ht.
	Dd / Mm		Registration No.		Time			Time											
	/																		
	/																		
	/																		
	/																		
	/																		
	/																		
	/																		
	/																		
	/																		

Totals this page:
Totals brought forward:
Totals to date:

I certify that the entries in this log are true.

_____

NOTES CONCERNING THE FLIGHTS LOGGED ON THE PAGE ABOVE

Flt. No.	Batteries / Fuel	P-f. Ins. / C. Cal.	Technical Problems Found / Corrective Action, Status	Remarks and Endorsements

FLT. NO.	DATE		AIRCRAFT		FLIGHT LOCATION			FLIGHT DURATION				FUNCTION			DATA			
	Yr ___		Make/Model		Takeoff	/	Landing	Fixed Wing	Rotary	LTA	Total	OIC/Solo	Training Received	Instructor	Night	Ldgs.	Dist.	Max. Ht.
	Dd / Mm		Registration No.		Time		Time											
	_/_																	
	_/_																	
	_/_																	
	_/_																	
	_/_																	
	_/_																	
	_/_																	
	_/_																	
	_/_																	
	I certify that the entries in this log are true.						Totals this page:											
	_____						Totals brought forward:											
							Totals to date:											

NOTES CONCERNING THE FLIGHTS LOGGED ON THE PAGE ABOVE

Flt. No.	Batteries / Fuel	P-f. Ins. / C. Cal.	Technical Problems Found / Corrective Action, Status	Remarks and Endorsements

FLT. NO.	DATE		AIRCRAFT		FLIGHT LOCATION			FLIGHT DURATION				FUNCTION			DATA			
	Yr		Make/Model		Takeoff	/	Landing	Fixed Wing	Rotary	LTA	Total	OIC/Solo	Training Received	Instructor	Night	Ldgs.	Dist.	Max. Ht.
	Dd / Mm		Registration No.		Time		Time											
	/																	
	/																	
	/																	
	/																	
	/																	
	/																	
	/																	
	/																	

I certify that the entries in this log are true.

_____

Totals this page:

Totals brought forward:

Totals to date:

NOTES CONCERNING THE FLIGHTS LOGGED ON THE PAGE ABOVE

Flt. No.	Batteries / Fuel	P-f. Ins. / C. Cal.	Technical Problems Found / Corrective Action, Status	Remarks and Endorsements

FLT. NO.	DATE		AIRCRAFT		FLIGHT LOCATION			FLIGHT DURATION				FUNCTION			DATA			
	Yr___	Dd / Mm	Make/Model	Registration No.	Takeoff Time	/	Landing Time	Fixed Wing	Rotary	LTA	Total	OIC/Solo	Training Received	Instructor	Night	Ldgs.	Dist.	Max. Ht.
	__/__																	
	__/__																	
	__/__																	
	__/__																	
	__/__																	
	__/__																	
	__/__																	
	__/__																	

I certify that the entries in this log are true.

_____

Totals this page:
Totals brought forward:
Totals to date:

Flt. No.	Batteries / Fuel	P-f. Ins. – – – C. Cal.	Technical Problems Found / Corrective Action, Status	Remarks and Endorsements

NOTES CONCERNING THE FLIGHTS LOGGED ON THE PAGE ABOVE

FLT. NO.	DATE		AIRCRAFT		FLIGHT LOCATION			FLIGHT DURATION				FUNCTION			DATA			
	Yr ___		Make/Model		Takeoff	/	Landing	Fixed Wing	Rotary	LTA	Total	OIC/Solo	Training Received	Instructor	Night	Ldgs.	Dist.	Max. Ht.
	Dd / Mm		Registration No.		Time		Time											
	_/_																	
	_/_																	
	_/_																	
	_/_																	
	_/_																	
	_/_																	
	_/_																	
	_/_																	
	_/_																	
	I certify that the entries in this log are true. _____							Totals this page:										
								Totals brought forward:										
								Totals to date:										

Flt. No.	Batteries / Fuel	P-f. Ins. / C. Cal.	NOTES CONCERNING THE FLIGHTS LOGGED ON THE PAGE ABOVE	
			Technical Problems Found / Corrective Action, Status	Remarks and Endorsements

FLT. NO.	DATE		AIRCRAFT		FLIGHT LOCATION			FLIGHT DURATION				FUNCTION			DATA			
	Yr ___		Make/Model		Takeoff	/	Landing	Fixed Wing	Rotary	LTA	Total	OIC/Solo	Training Received	Instructor	Night	Ldgs.	Dist.	Max. Ht.
	Dd / Mm		Registration No.		Time		Time											
	__/__																	
	__/__																	
	__/__																	
	__/__																	
	__/__																	
	__/__																	
	__/__																	
	__/__																	
							Totals this page:											
							Totals brought forward:											
	I certify that the entries in this log are true.						**Totals to date:**											

NOTES CONCERNING THE FLIGHTS LOGGED ON THE PAGE ABOVE

Flt. No.	Batteries / Fuel	P-f. Ins. / C. Cal.	Technical Problems Found / Corrective Action, Status	Remarks and Endorsements

FLT. NO.	DATE		AIRCRAFT		FLIGHT LOCATION			FLIGHT DURATION				FUNCTION			DATA			
	Yr___		Make/Model		Takeoff	/	Landing	Fixed Wing	Rotary	LTA	Total	OIC/Solo	Training Received	Instructor	Night	Ldgs.	Dist.	Max. Ht.
	Dd / Mm		Registration No.		Time		Time											
	/																	
	/																	
	/																	
	/																	
	/																	
	/																	
	/																	
	/																	

I certify that the entries in this log are true.

_____

Totals this page:
Totals brought forward:
Totals to date:

Flt. No.	Batteries / Fuel	P-f. Ins. --- C. Cal.	Technical Problems Found / Corrective Action, Status	Remarks and Endorsements

NOTES CONCERNING THE FLIGHTS LOGGED ON THE PAGE ABOVE

FLT. NO.	DATE		AIRCRAFT		FLIGHT LOCATION			FLIGHT DURATION				FUNCTION			DATA			
	Yr ___		Make/Model		Takeoff	/	Landing	Fixed Wing	Rotary	LTA	Total	OIC/Solo	Training Received	Instructor	Night	Ldgs.	Dist.	Max. Ht.
	Dd / Mm		Registration No.		Time		Time											
	_/_																	
	_/_																	
	_/_																	
	_/_																	
	_/_																	
	_/_																	
	_/_																	
	_/_																	
	_/_																	

I certify that the entries in this log are true.

Totals this page:

Totals brought forward:

Totals to date:

_____

Flt. No.	Batteries / Fuel	P-f. Ins. — — C. Cal.	Technical Problems Found / Corrective Action, Status	Remarks and Endorsements

NOTES CONCERNING THE FLIGHTS LOGGED ON THE PAGE ABOVE

FLT. NO.	DATE		AIRCRAFT		FLIGHT LOCATION			FLIGHT DURATION				FUNCTION			DATA			
	Yr___	Dd / Mm	Make/Model	Registration No.	Takeoff Time	/	Landing Time	Fixed Wing	Rotary	LTA	Total	OIC/Solo	Training Received	Instructor	Night	Ldgs.	Dist.	Max. Ht.
	_/_																	
	_/_																	
	_/_																	
	_/_																	
	_/_																	
	_/_																	
	_/_																	
	_/_																	
							Totals this page:											
							Totals brought forward:											
							Totals to date:											

I certify that the entries in this log are true.

_____

Flt. No.	Batteries / Fuel	P-f. Ins. / C. Cal.	Technical Problems Found / Corrective Action, Status	Remarks and Endorsements
			NOTES CONCERNING THE FLIGHTS LOGGED ON THE PAGE ABOVE	

FLT. No.	Date		Aircraft		Flight Location			Flight Duration				Function			Data			
	Yr ___		Make/Model		Takeoff	/	Landing	Fixed Wing	Rotary	LTA	Total	OIC/Solo	Training Received	Instructor	Night	Ldgs.	Dist.	Max. Ht.
	Dd / Mm		Registration No.		Time		Time											
	_/_																	
	_/_																	
	_/_																	
	_/_																	
	_/_																	
	_/_																	
	_/_																	
	_/_																	

I certify that the entries in this log are true.

_____

Totals this page:

Totals brought forward:

Totals to date:

NOTES CONCERNING THE FLIGHTS LOGGED ON THE PAGE ABOVE

Flt. No.	Batteries / Fuel	P-f. Ins. / C. Cal.	Technical Problems Found / Corrective Action, Status	Remarks and Endorsements

FLT. NO.	DATE		AIRCRAFT		FLIGHT LOCATION			FLIGHT DURATION				FUNCTION			DATA				
	Yr ___		Make/Model		Takeoff	/	Landing	Fixed Wing	Rotary	LTA	Total	OIC/Solo	Training Received	Instructor	Night	Ldgs.	Dist.	Max. Ht.
	Dd / Mm		Registration No.		Time		Time											
	/																	
	/																	
	/																	
	/																	
	/																	
	/																	
	/																	
	/																	
	/																	
							Totals this page:											
							Totals brought forward:											
							Totals to date:											

I certify that the entries in this log are true.

_____

142

Flt. No.	Batteries / Fuel	P-f. Ins. / C. Cal.	Technical Problems Found / Corrective Action, Status	Remarks and Endorsements

NOTES CONCERNING THE FLIGHTS LOGGED ON THE PAGE ABOVE

FLT. NO.	DATE		AIRCRAFT		FLIGHT LOCATION			FLIGHT DURATION				FUNCTION			DATA			
	Yr ___	Dd / Mm	Make/Model	Registration No.	Takeoff Time	/	Landing Time	Fixed Wing	Rotary	LTA	Total	OIC/Solo	Training Received	Instructor	Night	Ldgs.	Dist.	Max. Ht.
	_/_																	
	_/_																	
	_/_																	
	_/_																	
	_/_																	
	_/_																	
	_/_																	
	_/_																	

I certify that the entries in this log are true.

Totals this page:

Totals brought forward:

Totals to date:

NOTES CONCERNING THE FLIGHTS LOGGED ON THE PAGE ABOVE

Flt. No.	Batteries / Fuel	P-f. Ins. / C. Cal.	Technical Problems Found / Corrective Action, Status	Remarks and Endorsements

FLT. NO.	DATE		AIRCRAFT		FLIGHT LOCATION			FLIGHT DURATION				FUNCTION			DATA			
	Yr		Make/Model		Takeoff	/	Landing	Fixed Wing	Rotary	LTA	Total	OIC/Solo	Training Received	Instructor	Night	Ldgs.	Dist.	Max. Ht.
	Dd / Mm		Registration No.		Time		Time											
	_/_																	
	_/_																	
	_/_																	
	_/_																	
	_/_																	
	_/_																	
	_/_																	
	_/_																	
	_/_																	

Totals this page:

Totals brought forward:

Totals to date:

I certify that the entries in this log are true.

_____

Flt. No.	Batteries / Fuel	P-f. Ins. / C. Cal.	NOTES CONCERNING THE FLIGHTS LOGGED ON THE PAGE ABOVE	
			Technical Problems Found / Corrective Action, Status	Remarks and Endorsements

FLT. NO.	DATE		AIRCRAFT		FLIGHT LOCATION			FLIGHT DURATION				FUNCTION			DATA			
	Yr __	Dd / Mm	Make/Model	Registration No.	Takeoff Time	/	Landing Time	Fixed Wing	Rotary	LTA	Total	OIC/Solo	Training Received	Instructor	Night	Ldgs.	Dist.	Max. Ht.
	__/__																	
	__/__																	
	__/__																	
	__/__																	
	__/__																	
	__/__																	
	__/__																	
	__/__																	
	__/__																	

I certify that the entries in this log are true.

_____

Totals this page:

Totals brought forward:

Totals to date:

Flt. No.	Batteries / Fuel	P-f. Ins. — — C. Cal.	Technical Problems Found / Corrective Action, Status	Remarks and Endorsements

NOTES CONCERNING THE FLIGHTS LOGGED ON THE PAGE ABOVE

FLT. NO.	DATE		AIRCRAFT		FLIGHT LOCATION			FLIGHT DURATION				FUNCTION			DATA			
	Yr ___		Make/Model		Takeoff	/	Landing	Fixed Wing	Rotary	LTA	Total	OIC/Solo	Training Received	Instructor	Night	Ldgs.	Dist.	Max. Ht.
	Dd / Mm		Registration No.		Time		Time											
	_/_																	
	_/_																	
	_/_																	
	_/_																	
	_/_																	
	_/_																	
	_/_																	
	_/_																	
	_/_																	
	I certify that the entries in this log are true.						Totals this page:											
							Totals brought forward:											
	_____						**Totals to date:**											

150

Flt. No.	Batteries / Fuel	P-f. Ins. — — C. Cal.	Technical Problems Found / Corrective Action, Status	Remarks and Endorsements

NOTES CONCERNING THE FLIGHTS LOGGED ON THE PAGE ABOVE

FLT. No.	Date		Aircraft		Flight Location			Flight Duration				Function			Data			
	Yr ___		Make/Model		Takeoff	/	Landing	Fixed Wing	Rotary	LTA	Total	OIC/Solo	Training Received	Instructor	Night	Ldgs.	Dist.	Max. Ht.
	Dd / Mm		Registration No.		Time		Time											
	___/___																	
	___/___																	
	___/___																	
	___/___																	
	___/___																	
	___/___																	
	___/___																	
	___/___																	
	___/___																	
	I certify that the entries in this log are true.							Totals this page:										
	_____							Totals brought forward:										
								Totals to date:										

NOTES CONCERNING THE FLIGHTS LOGGED ON THE PAGE ABOVE

Flt. No.	Batteries / Fuel	P-f. Ins. / C. Cal.	Technical Problems Found / Corrective Action, Status	Remarks and Endorsements

FLT. NO.	DATE		AIRCRAFT		FLIGHT LOCATION			FLIGHT DURATION				FUNCTION			DATA			
	Yr ___		Make/Model		Takeoff	/	Landing	Fixed Wing	Rotary	LTA	Total	OIC/Solo	Training Received	Instructor	Night	Ldgs.	Dist.	Max. Ht.
	Dd / Mm		Registration No.		Time		Time											
	_/_																	
	_/_																	
	_/_																	
	_/_																	
	_/_																	
	_/_																	
	_/_																	
	_/_																	
							Totals this page:											
							Totals brought forward:											
I certify that the entries in this log are true. _____								**Totals to date:**										

Flt. No.	Batteries / Fuel	P-f. Ins. — C. Cal.	Technical Problems Found / Corrective Action, Status	Remarks and Endorsements

NOTES CONCERNING THE FLIGHTS LOGGED ON THE PAGE ABOVE

FLT. NO.	DATE		AIRCRAFT		FLIGHT LOCATION		FLIGHT DURATION				FUNCTION			DATA			
	Yr ___	Dd / Mm	Make/Model	Registration No.	Takeoff Time	Landing Time	Fixed Wing	Rotary	LTA	Total	OIC/Solo	Training Received	Instructor	Night	Ldgs.	Dist.	Max. Ht.
	_/_				/												
	_/_				/												
	_/_				/												
	_/_				/												
	_/_				/												
	_/_				/												
	_/_				/												
	_/_				/												
	_/_				/												

I certify that the entries in this log are true.

_____

Totals this page:

Totals brought forward:

Totals to date:

NOTES CONCERNING THE FLIGHTS LOGGED ON THE PAGE ABOVE

Flt. No.	Batteries / Fuel	P-f. Ins. / C. Cal.	Technical Problems Found / Corrective Action, Status	Remarks and Endorsements

Flt. No.	Date		Aircraft		Flight Location			Flight Duration				Function			Data			
	Yr ___	Dd / Mm	Make/Model	Registration No.	Takeoff Time	/	Landing Time	Fixed Wing	Rotary	LTA	Total	OIC/Solo	Training Received	Instructor	Night	Ldgs.	Dist.	Max. Ht.
		_/_																
		_/_																
		_/_																
		_/_																
		_/_																
		_/_																
		_/_																
		_/_																
		_/_																
					Totals this page:													
					Totals brought forward:													
					Totals to date:													

I certify that the entries in this log are true.

_____

NOTES CONCERNING THE FLIGHTS LOGGED ON THE PAGE ABOVE

Flt. No.	Batteries / Fuel	P-f. Ins. / C. Cal.	Technical Problems Found / Corrective Action, Status	Remarks and Endorsements

FLT. NO.	DATE Yr ___ Dd / Mm	AIRCRAFT Make/Model Registration No.	FLIGHT LOCATION Takeoff Time	/	Landing Time	FLIGHT DURATION Fixed Wing	Rotary	LTA	Total	FUNCTION OIC/Solo	Training Received	Instructor	DATA Night	Ldgs.	Dist.	Max. Ht.
	_/_															
	_/_															
	_/_															
	_/_															
	_/_															
	_/_															
	_/_															
	_/_															

I certify that the entries in this log are true.

_____

Totals this page:
Totals brought forward:
Totals to date:

Flt. No.	Batteries / Fuel	P-f. Ins. / C. Cal.	Technical Problems Found / Corrective Action, Status	Remarks and Endorsements

NOTES CONCERNING THE FLIGHTS LOGGED ON THE PAGE ABOVE

FLT. NO.	DATE		AIRCRAFT		FLIGHT LOCATION			FLIGHT DURATION				FUNCTION			DATA			
	Yr ___		Make/Model		Takeoff	/	Landing	Fixed Wing	Rotary	LTA	Total	OIC/Solo	Training Received	Instructor	Night	Ldgs.	Dist.	Max. Ht.
	Dd / Mm		Registration No.		Time		Time											
	/					/												
	/					/												
	/					/												
	/					/												
	/					/												
	/					/												
	/					/												
	/					/												
	/					/												

I certify that the entries in this log are true.

_____

Totals this page:

Totals brought forward:

Totals to date:

NOTES CONCERNING THE FLIGHTS LOGGED ON THE PAGE ABOVE

Flt. No.	Batteries / Fuel	P-f. Ins. / C. Cal.	Technical Problems Found / Corrective Action, Status	Remarks and Endorsements

FLT. NO.	DATE		AIRCRAFT		FLIGHT LOCATION			FLIGHT DURATION				FUNCTION			DATA			
	Yr ___	Dd / Mm	Make/Model	Registration No.	Takeoff Time	/	Landing Time	Fixed Wing	Rotary	LTA	Total	OIC/Solo	Training Received	Instructor	Night	Ldgs.	Dist.	Max. Ht.
	_/_																	
	_/_																	
	_/_																	
	_/_																	
	_/_																	
	_/_																	
	_/_																	
	_/_																	
							Totals this page:											
							Totals brought forward:											
							Totals to date:											

I certify that the entries in this log are true.

_____

NOTES CONCERNING THE FLIGHTS LOGGED ON THE PAGE ABOVE

Flt. No.	Batteries / Fuel	P-f. Ins. / C. Cal.	Technical Problems Found / Corrective Action, Status	Remarks and Endorsements

FLT. NO.	DATE		AIRCRAFT		FLIGHT LOCATION			FLIGHT DURATION				FUNCTION			DATA			
	Yr ___		Make/Model		Takeoff	/	Landing	Fixed Wing	Rotary	LTA	Total	OIC/Solo	Training Received	Instructor	Night	Ldgs.	Dist.	Max. Ht.
	Dd / Mm		Registration No.		Time		Time											
	___/___																	
	___/___																	
	___/___																	
	___/___																	
	___/___																	
	___/___																	
	___/___																	
	___/___																	
	___/___																	
	I certify that the entries in this log are true. _____							Totals this page:										
								Totals brought forward:										
								Totals to date:										

NOTES CONCERNING THE FLIGHTS LOGGED ON THE PAGE ABOVE

Flt. No.	Batteries / Fuel	P-f. Ins. / C. Cal.	Technical Problems Found / Corrective Action, Status	Remarks and Endorsements

FLT. No.	DATE Yr ___ Dd / Mm	AIRCRAFT Make/Model / Registration No.	FLIGHT LOCATION Takeoff Time	/	Landing Time	FLIGHT DURATION Fixed Wing	Rotary	LTA	Total	FUNCTION OIC/Solo	Training Received	Instructor	DATA Night	Ldgs.	Dist.	Max. Ht.
	___/___			/												
	___/___			/												
	___/___			/												
	___/___			/												
	___/___			/												
	___/___			/												
	___/___			/												
	___/___			/												
					Totals this page:											
					Totals brought forward:											
I certify that the entries in this log are true. _____					**Totals to date:**											

NOTES CONCERNING THE FLIGHTS LOGGED ON THE PAGE ABOVE

Flt. No.	Batteries / Fuel	P-f. Ins. – – C. Cal.	Technical Problems Found / Corrective Action, Status	Remarks and Endorsements

FLT. NO.	DATE		AIRCRAFT		FLIGHT LOCATION			FLIGHT DURATION				FUNCTION			DATA			
	Yr ___	Dd / Mm	Make/Model	Registration No.	Takeoff Time	/	Landing Time	Fixed Wing	Rotary	LTA	Total	OIC/Solo	Training Received	Instructor	Night	Ldgs.	Dist.	Max. Ht.
	/																	
	/																	
	/																	
	/																	
	/																	
	/																	
	/																	
	/																	
	/																	
											Totals this page:							
											Totals brought forward:							
	I certify that the entries in this log are true.										**Totals to date:**							

Flt. No.	Batteries / Fuel	P-f. Ins. – – C. Cal.	Technical Problems Found / Corrective Action, Status	Remarks and Endorsements

NOTES CONCERNING THE FLIGHTS LOGGED ON THE PAGE ABOVE

FLT. NO.	DATE	AIRCRAFT		FLIGHT LOCATION			FLIGHT DURATION				FUNCTION			DATA			
	Yr ___ Dd / Mm	Make/Model	Registration No.	Takeoff Time	/	Landing Time	Fixed Wing	Rotary	LTA	Total	OIC/Solo	Training Received	Instructor	Night	Ldgs.	Dist.	Max. Ht.
	___/___																
	___/___																
	___/___																
	___/___																
	___/___																
	___/___																
	___/___																
	___/___																

I certify that the entries in this log are true.

_____

Totals this page:

Totals brought forward:

Totals to date:

Flt. No.	Batteries / Fuel	P-f. Ins. / C. Cal.	NOTES CONCERNING THE FLIGHTS LOGGED ON THE PAGE ABOVE	
			Technical Problems Found / Corrective Action, Status	Remarks and Endorsements

FLT. NO.	DATE		AIRCRAFT		FLIGHT LOCATION			FLIGHT DURATION				FUNCTION			DATA			
	Yr___		Make/Model		Takeoff	/	Landing	Fixed Wing	Rotary	LTA	Total	OIC/Solo	Training Received	Instructor	Night	Ldgs.	Dist.	Max. Ht.
	Dd / Mm		Registration No.		Time		Time											
	/																	
	/																	
	/																	
	/																	
	/																	
	/																	
	/																	
	/																	
															Totals this page:			
															Totals brought forward:			
	I certify that the entries in this log are true.														**Totals to date:**			

NOTES CONCERNING THE FLIGHTS LOGGED ON THE PAGE ABOVE

Flt. No.	Batteries / Fuel	P-f. Ins. / C. Cal.	Technical Problems Found / Corrective Action, Status	Remarks and Endorsements

FLT. NO.	DATE		AIRCRAFT		FLIGHT LOCATION			FLIGHT DURATION				FUNCTION			DATA			
	Yr ___		Make/Model		Takeoff	/	Landing	Fixed Wing	Rotary	LTA	Total	OIC/Solo	Training Received	Instructor	Night	Ldgs.	Dist.	Max. Ht.
	Dd / Mm		Registration No.		Time		Time											
	_/_																	
	_/_																	
	_/_																	
	_/_																	
	_/_																	
	_/_																	
	_/_																	
	_/_																	
	I certify that the entries in this log are true. _____						Totals this page:											
							Totals brought forward:											
							Totals to date:											

NOTES CONCERNING THE FLIGHTS LOGGED ON THE PAGE ABOVE

Flt. No.	Batteries / Fuel	P-f. Ins. / C. Cal.	Technical Problems Found / Corrective Action, Status	Remarks and Endorsements

FLT. NO.	DATE		AIRCRAFT		FLIGHT LOCATION			FLIGHT DURATION				FUNCTION			DATA			
	Yr ___		Make/Model	Takeoff	/	Landing	Fixed Wing	Rotary	LTA	Total	OIC/Solo	Training Received	Instructor	Night	Ldgs.	Dist.	Max. Ht.
	Dd / Mm		Registration No.	Time		Time											
	/																
	/																
	/																
	/																
	/																
	/																
	/																
	/																
	/																

Totals this page:
Totals brought forward:
Totals to date:

I certify that the entries in this log are true.

_____

Flt. No.	Batteries / Fuel	P-f. Ins. --- C. Cal.	NOTES CONCERNING THE FLIGHTS LOGGED ON THE PAGE ABOVE	
			Technical Problems Found / Corrective Action, Status	Remarks and Endorsements

FLT. NO.	DATE		AIRCRAFT		FLIGHT LOCATION			FLIGHT DURATION				FUNCTION			DATA			
	Yr ___	Dd / Mm	Make/Model	Registration No.	Takeoff Time	/	Landing Time	Fixed Wing	Rotary	LTA	Total	OIC/Solo	Training Received	Instructor	Night	Ldgs.	Dist.	Max. Ht.
	__/__					/												
	__/__					/												
	__/__					/												
	__/__					/												
	__/__					/												
	__/__					/												
	__/__					/												
	__/__					/												
I certify that the entries in this log are true.					Totals this page:													
					Totals brought forward:													
_____					**Totals to date:**													

			NOTES CONCERNING THE FLIGHTS LOGGED ON THE PAGE ABOVE	
Flt. No.	Batteries / Fuel	P-f. Ins. — — — C. Cal.	Technical Problems Found / Corrective Action, Status	Remarks and Endorsements

FLT. NO.	DATE		AIRCRAFT		FLIGHT LOCATION			FLIGHT DURATION				FUNCTION			DATA			
	Yr___	Dd / Mm	Make/Model	Registration No.	Takeoff Time	/	Landing Time	Fixed Wing	Rotary	LTA	Total	OIC/Solo	Training Received	Instructor	Night	Ldgs.	Dist.	Max. Ht.
	_/_																	
	_/_																	
	_/_																	
	_/_																	
	_/_																	
	_/_																	
	_/_																	
	_/_																	
	_/_																	

Totals this page:
Totals brought forward:
Totals to date:

I certify that the entries in this log are true.

NOTES CONCERNING THE FLIGHTS LOGGED ON THE PAGE ABOVE

Flt. No.	Batteries / Fuel	P-f. Ins. / C. Cal.	Technical Problems Found / Corrective Action, Status	Remarks and Endorsements

FLT. NO.	DATE	AIRCRAFT		FLIGHT LOCATION			FLIGHT DURATION				FUNCTION			DATA			
	Yr ___	Make/Model		Takeoff	/	Landing	Fixed Wing	Rotary	LTA	Total	OIC/Solo	Training Received	Instructor	Night	Ldgs.	Dist.	Max. Ht.
	Dd / Mm	Registration No.		Time		Time											
	_/_																
	_/_																
	_/_																
	_/_																
	_/_																
	_/_																
	_/_																
	_/_																
	_/_																

I certify that the entries in this log are true.

Totals this page:
Totals brought forward:
Totals to date:

Flt. No.	Batteries / Fuel	P-f. Ins. — — C. Cal.	Notes concerning the flights logged on the page above	
			Technical Problems Found / Corrective Action, Status	Remarks and Endorsements

FLT. NO.	DATE Yr___ Dd/Mm	AIRCRAFT Make/Model	AIRCRAFT Registration No.	FLIGHT LOCATION Takeoff / Landing		FLIGHT DURATION				FUNCTION			DATA			
				Takeoff Time	Landing Time	Fixed Wing	Rotary	LTA	Total	OIC/Solo	Training Received	Instructor	Night	Ldgs.	Dist.	Max. Ht.
	/															
	/															
	/															
	/															
	/															
	/															
	/															
	/															
					Totals this page:											
					Totals brought forward:											
					Totals to date:											

I certify that the entries in this log are true.

_____

Flt. No.	Batteries / Fuel	P-f. Ins. / C. Cal.	NOTES CONCERNING THE FLIGHTS LOGGED ON THE PAGE ABOVE	
			Technical Problems Found / Corrective Action, Status	Remarks and Endorsements

FLT. NO.	DATE Yr___ Dd/Mm	AIRCRAFT Make/Model / Registration No.	FLIGHT LOCATION Takeoff Time / Landing Time	FLIGHT DURATION Fixed Wing	Rotary	LTA	Total	FUNCTION OIC/Solo	Training Received	Instructor	DATA Night	Ldgs.	Dist.	Max. Ht.
	_/_													
	_/_													
	_/_													
	_/_													
	_/_													
	_/_													
	_/_													
	_/_													
			Totals this page:											
			Totals brought forward:											
			Totals to date:											

I certify that the entries in this log are true.

_____

NOTES CONCERNING THE FLIGHTS LOGGED ON THE PAGE ABOVE

Flt. No.	Batteries / Fuel	P-f. Ins. / C. Cal.	Technical Problems Found / Corrective Action, Status	Remarks and Endorsements

FLT. NO.	DATE		AIRCRAFT		FLIGHT LOCATION			FLIGHT DURATION				FUNCTION			DATA			
	Yr___	Dd/Mm	Make/Model	Registration No.	Takeoff Time	/	Landing Time	Fixed Wing	Rotary	LTA	Total	OIC/Solo	Training Received	Instructor	Night	Ldgs.	Dist.	Max. Ht.
	_/_																	
	_/_																	
	_/_																	
	_/_																	
	_/_																	
	_/_																	
	_/_																	
	_/_																	
	_/_																	

Totals this page:
Totals brought forward:
Totals to date:

I certify that the entries in this log are true.

_____

NOTES CONCERNING THE FLIGHTS LOGGED ON THE PAGE ABOVE

Flt. No.	Batteries / Fuel	P-f. Ins. / C. Cal.	Technical Problems Found / Corrective Action, Status	Remarks and Endorsements

FLT. NO.	DATE		AIRCRAFT		FLIGHT LOCATION			FLIGHT DURATION				FUNCTION			DATA			
	Yr ___		Make/Model		Takeoff	/	Landing	Fixed Wing	Rotary	LTA	Total	OIC/Solo	Training Received	Instructor	Night	Ldgs.	Dist.	Max. Ht.
	Dd / Mm		Registration No.		Time		Time											
	_/_																	
	_/_																	
	_/_																	
	_/_																	
	_/_																	
	_/_																	
	_/_																	
	_/_																	

I certify that the entries in this log are true.

Totals this page:
Totals brought forward:
Totals to date:

NOTES CONCERNING THE FLIGHTS LOGGED ON THE PAGE ABOVE

Flt. No.	Batteries / Fuel	P-f. Ins. – – – C. Cal.	Technical Problems Found / Corrective Action, Status	Remarks and Endorsements

FLT. NO.	DATE		AIRCRAFT		FLIGHT LOCATION			FLIGHT DURATION				FUNCTION			DATA			
	Yr ___		Make/Model		Takeoff	/	Landing	Fixed Wing	Rotary	LTA	Total	OIC/Solo	Training Received	Instructor	Night	Ldgs.	Dist.	Max. Ht.
	Dd / Mm		Registration No.		Time		Time											
	/																	
	/																	
	/																	
	/																	
	/																	
	/																	
	/																	
	/																	
	/																	

I certify that the entries in this log are true.

_____

Totals this page:

Totals brought forward:

Totals to date:

Flt. No.	Batteries / Fuel	P-f. Ins. — — C. Cal.	Technical Problems Found / Corrective Action, Status	Remarks and Endorsements

NOTES CONCERNING THE FLIGHTS LOGGED ON THE PAGE ABOVE

FLT. NO.	DATE		AIRCRAFT		FLIGHT LOCATION			FLIGHT DURATION				FUNCTION			DATA			
	Yr ___	Dd / Mm	Make/Model	Registration No.	Takeoff Time	/ Landing Time		Fixed Wing	Rotary	LTA	Total	OIC/Solo	Training Received	Instructor	Night	Ldgs.	Dist.	Max. Ht.
	_/_																	
	_/_																	
	_/_																	
	_/_																	
	_/_																	
	_/_																	
	_/_																	
	_/_																	
							Totals this page:											
							Totals brought forward:											
							Totals to date:											

I certify that the entries in this log are true.

_____

Flt. No.	Batteries / Fuel	P-f. Ins. – – – C. Cal.	Technical Problems Found / Corrective Action, Status	Remarks and Endorsements
			NOTES CONCERNING THE FLIGHTS LOGGED ON THE PAGE ABOVE	

FLT. NO.	DATE		AIRCRAFT		FLIGHT LOCATION			FLIGHT DURATION				FUNCTION			DATA			
	Yr ___		Make/Model	Takeoff	/	Landing	Fixed Wing	Rotary	LTA	Total	OIC/Solo	Training Received	Instructor	Night	Ldgs.	Dist.	Max. Ht.
	Dd / Mm		Registration No.	Time		Time											
	_/_																
	_/_																
	_/_																
	_/_																
	_/_																
	_/_																
	_/_																
	_/_																
	_/_																

I certify that the entries in this log are true.

_____

Totals this page:

Totals brought forward:

Totals to date:

Flt. No.	Batteries / Fuel	P-f. Ins. / C. Cal.	Technical Problems Found / Corrective Action, Status	Remarks and Endorsements
			NOTES CONCERNING THE FLIGHTS LOGGED ON THE PAGE ABOVE	

FLT. NO.	DATE		AIRCRAFT		FLIGHT LOCATION		FLIGHT DURATION				FUNCTION			DATA			
	Yr ___	Dd / Mm	Make/Model	Registration No.	Takeoff Time	Landing / Time	Fixed Wing	Rotary	LTA	Total	OIC/Solo	Training Received	Instructor	Night	Ldgs.	Dist.	Max. Ht.
	_ / _																
	_ / _																
	_ / _																
	_ / _																
	_ / _																
	_ / _																
	_ / _																
	_ / _																

I certify that the entries in this log are true.

Totals this page:
Totals brought forward:
Totals to date:

NOTES CONCERNING THE FLIGHTS LOGGED ON THE PAGE ABOVE

Flt. No.	Batteries / Fuel	P-f. Ins. / C. Cal.	Technical Problems Found / Corrective Action, Status	Remarks and Endorsements

FLT. NO.	DATE		AIRCRAFT		FLIGHT LOCATION			FLIGHT DURATION				FUNCTION			DATA			
	Yr ___		Make/Model		Takeoff	/	Landing	Fixed Wing	Rotary	LTA	Total	OIC/Solo	Training Received	Instructor	Night	Ldgs.	Dist.	Max. Ht.
	Dd / Mm		Registration No.		Time		Time											
	___/___																	
	___/___																	
	___/___																	
	___/___																	
	___/___																	
	___/___																	
	___/___																	
	___/___																	
	___/___																	
							Totals this page:											
							Totals brought forward:											
							Totals to date:											

I certify that the entries in this log are true.

NOTES CONCERNING THE FLIGHTS LOGGED ON THE PAGE ABOVE

Flt. No.	Batteries / Fuel	P-f. Ins. / C. Cal.	Technical Problems Found / Corrective Action, Status	Remarks and Endorsements

FLT. NO.	DATE		AIRCRAFT		FLIGHT LOCATION			FLIGHT DURATION				FUNCTION			DATA			
	Yr___		Make/Model		Takeoff	/	Landing	Fixed Wing	Rotary	LTA	Total	OIC/Solo	Training Received	Instructor	Night	Ldgs.	Dist.	Max. Ht.
	Dd / Mm		Registration No.		Time		Time											
	_/_																	
	_/_																	
	_/_																	
	_/_																	
	_/_																	
	_/_																	
	_/_																	
	_/_																	
	_/_																	
											Totals this page:							
											Totals brought forward:							
											Totals to date:							

I certify that the entries in this log are true.

_____

Flt. No.	Batteries / Fuel	P-f. Ins. — — C. Cal.	Technical Problems Found / Corrective Action, Status	Remarks and Endorsements

NOTES CONCERNING THE FLIGHTS LOGGED ON THE PAGE ABOVE

FLT. NO.	DATE		AIRCRAFT		FLIGHT LOCATION			FLIGHT DURATION				FUNCTION			DATA			
	Yr___	Dd / Mm	Make/Model	Registration No.	Takeoff Time	/	Landing Time	Fixed Wing	Rotary	LTA	Total	OIC/Solo	Training Received	Instructor	Night	Ldgs.	Dist.	Max. Ht.
		_/_																
		_/_																
		_/_																
		_/_																
		_/_																
		_/_																
		_/_																
		_/_																
		_/_																

Totals this page:
Totals brought forward:
Totals to date:

I certify that the entries in this log are true.

Flt. No.	Batteries / Fuel	P-f. Ins. / C. Cal.	Technical Problems Found / Corrective Action, Status	Remarks and Endorsements

NOTES CONCERNING THE FLIGHTS LOGGED ON THE PAGE ABOVE

FLT. NO.	DATE Yr___ Dd/Mm	AIRCRAFT Make/Model / Registration No.	FLIGHT LOCATION Takeoff Time / Landing Time	FLIGHT DURATION Fixed Wing	Rotary	LTA	Total	FUNCTION OIC/Solo	Training Received	Instructor	DATA Night	Ldgs.	Dist.	Max. Ht.
	_/_													
	_/_													
	_/_													
	_/_													
	_/_													
	_/_													
	_/_													
	_/_													
	_/_													

Totals this page:
Totals brought forward:
Totals to date:

I certify that the entries in this log are true.

_____

NOTES CONCERNING THE FLIGHTS LOGGED ON THE PAGE ABOVE

Flt. No.	Batteries / Fuel	P-f. Ins. / C. Cal.	Technical Problems Found / Corrective Action, Status	Remarks and Endorsements

FLT. NO.	DATE		AIRCRAFT		FLIGHT LOCATION			FLIGHT DURATION				FUNCTION			DATA			
	Yr ___		Make/Model		Takeoff	/	Landing	Fixed Wing	Rotary	LTA	Total	OIC/Solo	Training Received	Instructor	Night	Ldgs.	Dist.	Max. Ht.
	Dd / Mm		Registration No.		Time		Time											
	/																	
	/																	
	/																	
	/																	
	/																	
	/																	
	/																	
	/																	
	/																	

Totals this page:
Totals brought forward:
Totals to date:

I certify that the entries in this log are true.

_____

NOTES CONCERNING THE FLIGHTS LOGGED ON THE PAGE ABOVE

Flt. No.	Batteries / Fuel	P-f. Ins. / C. Cal.	Technical Problems Found / Corrective Action, Status	Remarks and Endorsements

FLT. NO.	DATE		AIRCRAFT		FLIGHT LOCATION			FLIGHT DURATION				FUNCTION			DATA			
	Yr___		Make/Model		Takeoff	/	Landing	Fixed Wing	Rotary	LTA	Total	OIC/Solo	Training Received	Instructor	Night	Ldgs.	Dist.	Max. Ht.
	Dd / Mm		Registration No.		Time		Time											
	_/_																	
	_/_																	
	_/_																	
	_/_																	
	_/_																	
	_/_																	
	_/_																	
	_/_																	

I certify that the entries in this log are true.

Totals this page:

Totals brought forward:

Totals to date:

Flt. No.	Batteries / Fuel	P-f. Ins. / C. Cal.	Technical Problems Found / Corrective Action, Status	Remarks and Endorsements

NOTES CONCERNING THE FLIGHTS LOGGED ON THE PAGE ABOVE

FLT. NO.	DATE		AIRCRAFT		FLIGHT LOCATION			FLIGHT DURATION				FUNCTION			DATA			
	Yr ___		Make/Model		Takeoff	/	Landing	Fixed Wing	Rotary	LTA	Total	OIC/Solo	Training Received	Instructor	Night	Ldgs.	Dist.	Max. Ht.
	Dd / Mm		Registration No.		Time		Time											
	/					/												
	/					/												
	/					/												
	/					/												
	/					/												
	/					/												
	/					/												
	/					/												
	/					/												

Totals this page:

Totals brought forward:

Totals to date:

I certify that the entries in this log are true.

_____

NOTES CONCERNING THE FLIGHTS LOGGED ON THE PAGE ABOVE

Flt. No.	Batteries / Fuel	P-f. Ins. / C. Cal.	Technical Problems Found / Corrective Action, Status	Remarks and Endorsements

FLT. NO.	DATE		AIRCRAFT		FLIGHT LOCATION		FLIGHT DURATION				FUNCTION			DATA			
	Yr ___		Make/Model		Takeoff	/ Landing	Fixed Wing	Rotary	LTA	Total	OIC/Solo	Training Received	Instructor	Night	Ldgs.	Dist.	Max. Ht.
	Dd / Mm		Registration No.		Time	Time											
	_/_																
	_/_																
	_/_																
	_/_																
	_/_																
	_/_																
	_/_																
	_/_																
	I certify that the entries in this log are true.					Totals this page:											
						Totals brought forward:											
						Totals to date:											

			NOTES CONCERNING THE FLIGHTS LOGGED ON THE PAGE ABOVE	
Flt. No.	Batteries / Fuel	P-f. Ins. — — C. Cal.	Technical Problems Found / Corrective Action, Status	Remarks and Endorsements

FLT. NO.	DATE		AIRCRAFT		FLIGHT LOCATION			FLIGHT DURATION				FUNCTION			DATA			
	Yr ___	Dd / Mm	Make/Model	Registration No.	Takeoff Time	/	Landing Time	Fixed Wing	Rotary	LTA	Total	OIC/Solo	Training Received	Instructor	Night	Ldgs.	Dist.	Max. Ht.
	/					/												
	/					/												
	/					/												
	/					/												
	/					/												
	/					/												
	/					/												
	/					/												
	/					/												

I certify that the entries in this log are true.

_____

Totals this page:
Totals brought forward:
Totals to date:

Flt. No.	Batteries / Fuel	P-f. Ins. / C. Cal.	Technical Problems Found / Corrective Action, Status	Remarks and Endorsements

NOTES CONCERNING THE FLIGHTS LOGGED ON THE PAGE ABOVE

FLT. NO.	DATE		AIRCRAFT		FLIGHT LOCATION			FLIGHT DURATION				FUNCTION			DATA			
	Yr ___	Dd / Mm	Make/Model	Registration No.	Takeoff Time	/	Landing Time	Fixed Wing	Rotary	LTA	Total	OIC/Solo	Training Received	Instructor	Night	Ldgs.	Dist.	Max. Ht.
	_/_																	
	_/_																	
	_/_																	
	_/_																	
	_/_																	
	_/_																	
	_/_																	
	_/_																	
	_/_																	

I certify that the entries in this log are true.

_____

Totals this page:

Totals brought forward:

Totals to date:

Flt. No.	Batteries / Fuel	P-f. Ins. — — — C. Cal.	Technical Problems Found / Corrective Action, Status	Remarks and Endorsements

NOTES CONCERNING THE FLIGHTS LOGGED ON THE PAGE ABOVE

FLT. NO.	DATE		AIRCRAFT		FLIGHT LOCATION		FLIGHT DURATION				FUNCTION			DATA			
	Yr ___	Dd / Mm	Make/Model	Registration No.	Takeoff / Time	Landing Time	Fixed Wing	Rotary	LTA	Total	OIC/Solo	Training Received	Instructor	Night	Ldgs.	Dist.	Max. Ht.
	_/_																
	_/_																
	_/_																
	_/_																
	_/_																
	_/_																
	_/_																
	_/_																
	_/_																
										Totals this page:							
										Totals brought forward:							
I certify that the entries in this log are true. _____										**Totals to date:**							

Flt. No.	Batteries / Fuel	P-f. Ins. — — C. Cal.	Technical Problems Found / Corrective Action, Status	Remarks and Endorsements
			NOTES CONCERNING THE FLIGHTS LOGGED ON THE PAGE ABOVE	

FLT. NO.	DATE		AIRCRAFT		FLIGHT LOCATION			FLIGHT DURATION				FUNCTION			DATA			
	Yr ___		Make/Model		Takeoff	/	Landing	Fixed Wing	Rotary	LTA	Total	OIC/Solo	Training Received	Instructor	Night	Ldgs.	Dist.	Max. Ht.
	Dd / Mm		Registration No.		Time		Time											
	_/_																	
	_/_																	
	_/_																	
	_/_																	
	_/_																	
	_/_																	
	_/_																	
	_/_																	
	I certify that the entries in this log are true.						Totals this page:											
							Totals brought forward:											
							Totals to date:											

Flt. No.	Batteries / Fuel	P-f. Ins. / C. Cal.	Technical Problems Found / Corrective Action, Status	Remarks and Endorsements

NOTES CONCERNING THE FLIGHTS LOGGED ON THE PAGE ABOVE

FLT. NO.	DATE		AIRCRAFT		FLIGHT LOCATION			FLIGHT DURATION				FUNCTION			DATA			
	Yr ___		Make/Model		Takeoff	/	Landing	Fixed Wing	Rotary	LTA	Total	OIC/Solo	Training Received	Instructor	Night	Ldgs.	Dist.	Max. Ht.
	Dd / Mm		Registration No.		Time		Time											
	/																	
	/																	
	/																	
	/																	
	/																	
	/																	
	/																	
	/																	

I certify that the entries in this log are true.

_____

Totals this page:
Totals brought forward:
Totals to date:

Flt. No.	Batteries / Fuel	P-f. Ins. / C. Cal.	Technical Problems Found / Corrective Action, Status	Remarks and Endorsements

NOTES CONCERNING THE FLIGHTS LOGGED ON THE PAGE ABOVE

FLT. NO.	DATE		AIRCRAFT		FLIGHT LOCATION			FLIGHT DURATION				FUNCTION			DATA			
	Yr___	Dd / Mm	Make/Model	Registration No.	Takeoff Time	/	Landing Time	Fixed Wing	Rotary	LTA	Total	OIC/Solo	Training Received	Instructor	Night	Ldgs.	Dist.	Max. Ht.
	_ / _					/												
	_ / _					/												
	_ / _					/												
	_ / _					/												
	_ / _					/												
	_ / _					/												
	_ / _					/												
	_ / _					/												

I certify that the entries in this log are true.

Totals this page:

Totals brought forward:

Totals to date:

NOTES CONCERNING THE FLIGHTS LOGGED ON THE PAGE ABOVE				
Flt. No.	Batteries / Fuel	P-f. Ins. — — C. Cal.	Technical Problems Found / Corrective Action, Status	Remarks and Endorsements

FLT. NO.	DATE		AIRCRAFT		FLIGHT LOCATION		FLIGHT DURATION				FUNCTION			DATA			
	Yr____	Dd / Mm	Make/Model	Registration No.	Takeoff Time	/ Landing Time	Fixed Wing	Rotary	LTA	Total	OIC/Solo	Training Received	Instructor	Night	Ldgs.	Dist.	Max. Ht.
	/				/												
	/				/												
	/				/												
	/				/												
	/				/												
	/				/												
	/				/												
	/				/												
	/				/												
						Totals this page:											
						Totals brought forward:											
I certify that the entries in this log are true. _____						**Totals to date:**											

Flt. No.	Batteries / Fuel	P-f. Ins. – – – C. Cal.	Technical Problems Found / Corrective Action, Status	Remarks and Endorsements

NOTES CONCERNING THE FLIGHTS LOGGED ON THE PAGE ABOVE

FLT. NO.	DATE		AIRCRAFT		FLIGHT LOCATION			FLIGHT DURATION				FUNCTION			DATA			
	Yr ____	Dd / Mm	Make/Model	Registration No.	Takeoff Time	/	Landing Time	Fixed Wing	Rotary	LTA	Total	OIC/Solo	Training Received	Instructor	Night	Ldgs.	Dist.	Max. Ht.
	_/_																	
	_/_																	
	_/_																	
	_/_																	
	_/_																	
	_/_																	
	_/_																	
	_/_																	

I certify that the entries in this log are true.

Totals this page:

Totals brought forward:

Totals to date:

Flt. No.	Batteries / Fuel	P-f. Ins. —— C. Cal.	Notes concerning the flights logged on the page above	
			Technical Problems Found / Corrective Action, Status	Remarks and Endorsements

FLT. NO.	DATE		AIRCRAFT		FLIGHT LOCATION			FLIGHT DURATION				FUNCTION			DATA			
	Yr ___		Make/Model		Takeoff	/	Landing	Fixed Wing	Rotary	LTA	Total	OIC/Solo	Training Received	Instructor	Night	Ldgs.	Dist.	Max. Ht.
	Dd / Mm		Registration No.		Time		Time											
	_/_																	
	_/_																	
	_/_																	
	_/_																	
	_/_																	
	_/_																	
	_/_																	
	_/_																	

I certify that the entries in this log are true.

_____

Totals this page:
Totals brought forward:
Totals to date:

Flt. No.	Batteries / Fuel	P-f. Ins. — — C. Cal.	Technical Problems Found / Corrective Action, Status	Remarks and Endorsements

NOTES CONCERNING THE FLIGHTS LOGGED ON THE PAGE ABOVE

FLT. NO.	DATE Yr___ Dd/Mm	AIRCRAFT Make/Model / Registration No.	FLIGHT LOCATION Takeoff Time	Landing Time	FLIGHT DURATION Fixed Wing	Rotary	LTA	Total	FUNCTION OIC/Solo	Training Received	Instructor	DATA Night	Ldgs.	Dist.	Max. Ht.
	__/__														
	__/__														
	__/__														
	__/__														
	__/__														
	__/__														
	__/__														
	__/__														
				Totals this page:											
				Totals brought forward:											
I certify that the entries in this log are true. _____				**Totals to date:**											

Flt. No.	Batteries / Fuel	P-f. Ins. — C. Cal.	Technical Problems Found / Corrective Action, Status	Remarks and Endorsements

FLT. NO.	DATE		AIRCRAFT		FLIGHT LOCATION			FLIGHT DURATION				FUNCTION			DATA			
	Yr ___		Make/Model		Takeoff	/	Landing	Fixed Wing	Rotary	LTA	Total	OIC/Solo	Training Received	Instructor	Night	Ldgs.	Dist.	Max. Ht.
	Dd / Mm		Registration No.		Time		Time											
	_/_																	
	_/_																	
	_/_																	
	_/_																	
	_/_																	
	_/_																	
	_/_																	
	_/_																	
	_/_																	

I certify that the entries in this log are true.

_____

Totals this page:

Totals brought forward:

Totals to date:

NOTES CONCERNING THE FLIGHTS LOGGED ON THE PAGE ABOVE

Flt. No.	Batteries / Fuel	P-f. Ins. / C. Cal.	Technical Problems Found / Corrective Action, Status	Remarks and Endorsements

FLT. NO.	DATE Yr ___ Dd/Mm	AIRCRAFT Make/Model / Registration No.	FLIGHT LOCATION Takeoff Time / Landing Time	FLIGHT DURATION Fixed Wing	Rotary	LTA	Total	FUNCTION OIC/Solo	Training Received	Instructor	DATA Night	Ldgs.	Dist.	Max. Ht.
	_/_													
	_/_													
	_/_													
	_/_													
	_/_													
	_/_													
	_/_													
	_/_													
	_/_													
			Totals this page:											
I certify that the entries in this log are true. _____			Totals brought forward:											
			Totals to date:											

NOTES CONCERNING THE FLIGHTS LOGGED ON THE PAGE ABOVE

Flt. No.	Batteries / Fuel	P-f. Ins. – – C. Cal.	Technical Problems Found / Corrective Action, Status	Remarks and Endorsements

FLT. NO.	DATE		AIRCRAFT		FLIGHT LOCATION			FLIGHT DURATION				FUNCTION			DATA			
	Yr ___		Make/Model		Takeoff	/	Landing	Fixed Wing	Rotary	LTA	Total	OIC/Solo	Training Received	Instructor	Night	Ldgs.	Dist.	Max. Ht.
	Dd / Mm		Registration No.		Time		Time											
	___/___																	
	___/___																	
	___/___																	
	___/___																	
	___/___																	
	___/___																	
	___/___																	
	___/___																	
	___/___																	

I certify that the entries in this log are true.

_____

Totals this page:

Totals brought forward:

Totals to date:

242

NOTES CONCERNING THE FLIGHTS LOGGED ON THE PAGE ABOVE

Flt. No.	Batteries / Fuel	P-f. Ins. / C. Cal.	Technical Problems Found / Corrective Action, Status	Remarks and Endorsements

FLT. NO.	DATE		AIRCRAFT		FLIGHT LOCATION			FLIGHT DURATION				FUNCTION			DATA			
	Yr ___		Make/Model		Takeoff	/	Landing	Fixed Wing	Rotary	LTA	Total	OIC/Solo	Training Received	Instructor	Night	Ldgs.	Dist.	Max. Ht.
	Dd / Mm		Registration No.		Time		Time											
	/																	
	/																	
	/																	
	/																	
	/																	
	/																	
	/																	
	/																	
												Totals this page:						
	I certify that the entries in this log are true.											Totals brought forward:						
	_____											**Totals to date:**						

Notes concerning the flights logged on the page above

Flt. No.	Batteries / Fuel	P-f. Ins. / C. Cal.	Technical Problems Found / Corrective Action, Status	Remarks and Endorsements

FLT. NO.	DATE		AIRCRAFT		FLIGHT LOCATION				FLIGHT DURATION				FUNCTION			DATA			
	Yr ___		Make/Model		Takeoff	/	Landing		Fixed Wing	Rotary	LTA	Total	OIC/Solo	Training Received	Instructor	Night	Ldgs.	Dist.	Max. Ht.
	Dd / Mm		Registration No.		Time		Time												
	_/_																		
	_/_																		
	_/_																		
	_/_																		
	_/_																		
	_/_																		
	_/_																		
	_/_																		
	_/_																		
								Totals this page:											
	I certify that the entries in this log are true. _____								Totals brought forward:										
								Totals to date:											

246

Flt. No.	Batteries / Fuel	P-f. Ins. — — C. Cal.	NOTES CONCERNING THE FLIGHTS LOGGED ON THE PAGE ABOVE	
			Technical Problems Found / Corrective Action, Status	Remarks and Endorsements

FLT. NO.	DATE		AIRCRAFT		FLIGHT LOCATION			FLIGHT DURATION				FUNCTION			DATA			
	Yr___	Dd / Mm	Make/Model	Registration No.	Takeoff Time	/	Landing Time	Fixed Wing	Rotary	LTA	Total	OIC/Solo	Training Received	Instructor	Night	Ldgs.	Dist.	Max. Ht.
	_/_																	
	_/_																	
	_/_																	
	_/_																	
	_/_																	
	_/_																	
	_/_																	
	_/_																	
	_/_																	

I certify that the entries in this log are true.
_____

Totals this page:
Totals brought forward:
Totals to date:

			NOTES CONCERNING THE FLIGHTS LOGGED ON THE PAGE ABOVE	
Flt. No.	Batteries / Fuel	P-f. Ins. / C. Cal.	Technical Problems Found / Corrective Action, Status	Remarks and Endorsements

FLT. NO.	DATE		AIRCRAFT		FLIGHT LOCATION			FLIGHT DURATION				FUNCTION			DATA			
	Yr ___		Make/Model		Takeoff	/	Landing	Fixed Wing	Rotary	LTA	Total	OIC/Solo	Training Received	Instructor	Night	Ldgs.	Dist.	Max. Ht.
	Dd / Mm		Registration No.		Time		Time											
	__/__																	
	__/__																	
	__/__																	
	__/__																	
	__/__																	
	__/__																	
	__/__																	
	__/__																	
	__/__																	
							Totals this page:											
							Totals brought forward:											
							Totals to date:											

I certify that the entries in this log are true.

_____

Flt. No.	Batteries / Fuel	P-f. Ins. / C. Cal.	Technical Problems Found / Corrective Action, Status	Remarks and Endorsements

FLT. NO.	DATE Yr___ Dd/Mm	AIRCRAFT Make/Model / Registration No.	FLIGHT LOCATION Takeoff Time / Landing Time	FLIGHT DURATION Fixed Wing	Rotary	LTA	Total	FUNCTION OIC/Solo	Training Received	Instructor	DATA Night	Ldgs.	Dist.	Max. Ht.
	_/_													
	_/_													
	_/_													
	_/_													
	_/_													
	_/_													
	_/_													
	_/_													
	_/_													

I certify that the entries in this log are true.

_____

Totals this page:
Totals brought forward:
Totals to date:

NOTES CONCERNING THE FLIGHTS LOGGED ON THE PAGE ABOVE

Flt. No.	Batteries / Fuel	P-f. Ins. / C. Cal.	Technical Problems Found / Corrective Action, Status	Remarks and Endorsements

FLT. NO.	DATE		AIRCRAFT		FLIGHT LOCATION		FLIGHT DURATION				FUNCTION			DATA			
	Yr ___	Dd / Mm	Make/Model	Registration No.	Takeoff / Time	Landing Time	Fixed Wing	Rotary	LTA	Total	OIC/Solo	Training Received	Instructor	Night	Ldgs.	Dist.	Max. Ht.
	__/__																
	__/__																
	__/__																
	__/__																
	__/__																
	__/__																
	__/__																
	__/__																
	__/__																

Totals this page:
Totals brought forward:
Totals to date:

I certify that the entries in this log are true.

_____

NOTES CONCERNING THE FLIGHTS LOGGED ON THE PAGE ABOVE

Flt. No.	Batteries / Fuel	P-f. Ins. / C. Cal.	Technical Problems Found / Corrective Action, Status	Remarks and Endorsements

FLT. NO.	DATE		AIRCRAFT		FLIGHT LOCATION			FLIGHT DURATION				FUNCTION			DATA			
	Yr___	Dd / Mm	Make/Model	Registration No.	Takeoff Time	/	Landing Time	Fixed Wing	Rotary	LTA	Total	OIC/Solo	Training Received	Instructor	Night	Ldgs.	Dist.	Max. Ht.
	/																	
	/																	
	/																	
	/																	
	/																	
	/																	
	/																	
	/																	

I certify that the entries in this log are true.

_____

Totals this page:
Totals brought forward:
Totals to date:

NOTES CONCERNING THE FLIGHTS LOGGED ON THE PAGE ABOVE

Flt. No.	Batteries / Fuel	P-f. Ins. / C. Cal.	Technical Problems Found / Corrective Action, Status	Remarks and Endorsements

FLT. NO.	DATE		AIRCRAFT		FLIGHT LOCATION			FLIGHT DURATION				FUNCTION			DATA			
	Yr		Make/Model		Takeoff	/	Landing	Fixed Wing	Rotary	LTA	Total	OIC/Solo	Training Received	Instructor	Night	Ldgs.	Dist.	Max. Ht.
	Dd / Mm		Registration No.		Time		Time											
	/																	
	/																	
	/																	
	/																	
	/																	
	/																	
	/																	
	/																	

Totals this page:
Totals brought forward:
Totals to date:

I certify that the entries in this log are true.

NOTES CONCERNING THE FLIGHTS LOGGED ON THE PAGE ABOVE

Flt. No.	Batteries / Fuel	P-f. Ins. / C. Cal.	Technical Problems Found / Corrective Action, Status	Remarks and Endorsements

FLT. NO.	DATE Yr___ Dd/Mm	AIRCRAFT Make/Model / Registration No.	FLIGHT LOCATION Takeoff Time / Landing Time	FLIGHT DURATION Fixed Wing	Rotary	LTA	Total	FUNCTION OIC/Solo	Training Received	Instructor	DATA Night	Ldgs.	Dist.	Max. Ht.
	_/_		/											
	_/_		/											
	_/_		/											
	_/_		/											
	_/_		/											
	_/_		/											
	_/_		/											
	_/_		/											
			Totals this page:											
			Totals brought forward:											
			Totals to date:											

I certify that the entries in this log are true.

_____

Flt. No.	Batteries / Fuel	P-f. Ins. — — C. Cal.	Technical Problems Found / Corrective Action, Status	Remarks and Endorsements

FLT. NO.	DATE		AIRCRAFT		FLIGHT LOCATION			FLIGHT DURATION				FUNCTION			DATA			
	Yr ___		Make/Model		Takeoff	/	Landing	Fixed Wing	Rotary	LTA	Total	OIC/Solo	Training Received	Instructor	Night	Ldgs.	Dist.	Max. Ht.
	Dd / Mm		Registration No.		Time		Time											
	—/—																	
	—/—																	
	—/—																	
	—/—																	
	—/—																	
	—/—																	
	—/—																	
	—/—																	

Totals this page:

Totals brought forward:

Totals to date:

I certify that the entries in this log are true.

_____

			NOTES CONCERNING THE FLIGHTS LOGGED ON THE PAGE ABOVE	
Flt. No.	Batteries / Fuel	P-f. Ins. – – – C. Cal.	Technical Problems Found / Corrective Action, Status	Remarks and Endorsements

FLT. NO.	DATE		AIRCRAFT		FLIGHT LOCATION			FLIGHT DURATION				FUNCTION			DATA			
	Yr___	Dd / Mm	Make/Model	Registration No.	Takeoff Time	/	Landing Time	Fixed Wing	Rotary	LTA	Total	OIC/Solo	Training Received	Instructor	Night	Ldgs.	Dist.	Max. Ht.
	___/___																	
	___/___																	
	___/___																	
	___/___																	
	___/___																	
	___/___																	
	___/___																	
	___/___																	
	___/___																	
	I certify that the entries in this log are true. _____						Totals this page: Totals brought forward: **Totals to date:**											

Flt. No.	Batteries / Fuel	P-f. Ins. — — C. Cal.	Technical Problems Found / Corrective Action, Status	Remarks and Endorsements

NOTES CONCERNING THE FLIGHTS LOGGED ON THE PAGE ABOVE

FLT. NO.	DATE		AIRCRAFT		FLIGHT LOCATION			FLIGHT DURATION				FUNCTION			DATA			
	Yr ___	Dd / Mm	Make/Model	Registration No.	Takeoff Time	/	Landing Time	Fixed Wing	Rotary	LTA	Total	OIC/Solo	Training Received	Instructor	Night	Ldgs.	Dist.	Max. Ht.
	_/_																	
	_/_																	
	_/_																	
	_/_																	
	_/_																	
	_/_																	
	_/_																	
	_/_																	
						Totals this page:												
						Totals brought forward:												
	I certify that the entries in this log are true.					**Totals to date:**												

Flt. No.	Batteries / Fuel	P-f. Ins. / C. Cal.	Technical Problems Found / Corrective Action, Status	Remarks and Endorsements

FLT. NO.	DATE		AIRCRAFT		FLIGHT LOCATION			FLIGHT DURATION				FUNCTION			DATA			
	Yr ___		Make/Model		Takeoff	/	Landing	Fixed Wing	Rotary	LTA	Total	OIC/Solo	Training Received	Instructor	Night	Ldgs.	Dist.	Max. Ht.
	Dd / Mm		Registration No.		Time		Time											
	/					/												
	/					/												
	/					/												
	/					/												
	/					/												
	/					/												
	/					/												
	/					/												

I certify that the entries in this log are true.

Totals this page:

Totals brought forward:

Totals to date:

Flt. No.	Batteries / Fuel	P-f. Ins. / C. Cal.	Technical Problems Found / Corrective Action, Status	Remarks and Endorsements

FLT. NO.	DATE		AIRCRAFT		FLIGHT LOCATION				FLIGHT DURATION				FUNCTION			DATA			
	Yr ___		Make/Model		Takeoff	/	Landing		Fixed Wing	Rotary	LTA	Total	OIC/Solo	Training Received	Instructor	Night	Ldgs.	Dist.	Max. Ht.
	Dd / Mm		Registration No.		Time		Time												
	_/_																		
	_/_																		
	_/_																		
	_/_																		
	_/_																		
	_/_																		
	_/_																		
	_/_																		
	_/_																		

Totals this page:
Totals brought forward:
Totals to date:

I certify that the entries in this log are true.

Flt. No.	Batteries / Fuel	P-f. Ins. / C. Cal.	Technical Problems Found / Corrective Action, Status	Remarks and Endorsements

NOTES CONCERNING THE FLIGHTS LOGGED ON THE PAGE ABOVE

FLT. NO.	DATE		AIRCRAFT		FLIGHT LOCATION				FLIGHT DURATION				FUNCTION			DATA			
	Yr ___	Dd / Mm	Make/Model	Registration No.	Takeoff Time	/	Landing Time		Fixed Wing	Rotary	LTA	Total	OIC/Solo	Training Received	Instructor	Night	Ldgs.	Dist.	Max. Ht.
	/					/													
	/					/													
	/					/													
	/					/													
	/					/													
	/					/													
	/					/													
	/					/													

I certify that the entries in this log are true.

_____

Totals this page:
Totals brought forward:
Totals to date:

Flt. No.	Batteries / Fuel	P-f. Ins. / C. Cal.	Technical Problems Found / Corrective Action, Status	Remarks and Endorsements

NOTES CONCERNING THE FLIGHTS LOGGED ON THE PAGE ABOVE

FLT. NO.	DATE		AIRCRAFT		FLIGHT LOCATION		FLIGHT DURATION				FUNCTION			DATA			
	Yr ___		Make/Model		Takeoff	/ Landing	Fixed Wing	Rotary	LTA	Total	OIC/Solo	Training Received	Instructor	Night	Ldgs.	Dist.	Max. Ht.
	Dd / Mm		Registration No.		Time	Time											
	_/_																
	_/_																
	_/_																
	_/_																
	_/_																
	_/_																
	_/_																
	_/_																

Totals this page:

Totals brought forward:

Totals to date:

I certify that the entries in this log are true.

_____

Notes concerning the flights logged on the page above

Flt. No.	Batteries / Fuel	P-f. Ins. / C. Cal.	Technical Problems Found / Corrective Action, Status	Remarks and Endorsements

PER-AIRCRAFT TOTALS (16 aircraft is the maximum number identified in Section 1 of this logbook)

AIRCRAFT	FLIGHT TIME				FUNCTION			DATA	
Registration No.	Fixed	Rotary	LTA	Total	OIC/Solo	Training	Instructor	Night	Ldgs
1									
2									
3									
4									
5									
6									
7									
8									
9									
10									
11									
12									
13									
14									
15									
16									
Totals all aircraft (should equal totals on page 274):									

Part 3: Record of Pilot Training

Flight and Ground Training Record

Date	Course Title	Duration	Test Score	Pass Y/N	Instructor / Examiner Name, Signature, Certificate Number

Flight and Ground Training Record, continued

Date	Course Title	Duration	Test Score	Pass Y/N	Instructor / Examiner Name, Signature, Certificate Number

TRAINING CERTIFICATIONS

I certify that _____
has satisfactorily completed the aviation training course/program titled
_____

Signed _____ Date _____

Cert. # _____ Expiration _____

I certify that _____
has satisfactorily completed the aviation training course/program titled
_____

Signed _____ Date _____

Cert. # _____ Expiration _____

I certify that _____
has satisfactorily completed the aviation training course/program titled
_____

Signed _____ Date _____

Cert. # _____ Expiration _____

I certify that _____
has satisfactorily completed the aviation training course/program titled
_____

Signed _____ Date _____

Cert. # _____ Expiration _____

TRAINING CERTIFICATIONS, CONTINUED

I certify that _____
has satisfactorily completed the aviation training course/program titled
_____
_____

Signed _____ Date _____

Cert. # _____ Expiration _____

I certify that _____
has satisfactorily completed the aviation training course/program titled
_____
_____

Signed _____ Date _____

Cert. # _____ Expiration _____

I certify that _____
has satisfactorily completed the aviation training course/program titled
_____
_____

Signed _____ Date _____

Cert. # _____ Expiration _____

I certify that _____
has satisfactorily completed the aviation training course/program titled
_____
_____

Signed _____ Date _____

Cert. # _____ Expiration _____

TRAINING CERTIFICATIONS, CONTINUED

I certify that _____
has satisfactorily completed the aviation training course/program titled
_____
_____

Signed _____ Date _____

Cert. # _____ Expiration _____

I certify that _____
has satisfactorily completed the aviation training course/program titled
_____
_____

Signed _____ Date _____

Cert. # _____ Expiration _____

I certify that _____
has satisfactorily completed the aviation training course/program titled
_____
_____

Signed _____ Date _____

Cert. # _____ Expiration _____

I certify that _____
has satisfactorily completed the aviation training course/program titled
_____
_____

Signed _____ Date _____

Cert. # _____ Expiration _____

TRAINING CERTIFICATIONS, CONTINUED

I certify that _____
has satisfactorily completed the aviation training course/program titled

_____

Signed _____ Date _____

Cert. # _____ Expiration _____

I certify that _____
has satisfactorily completed the aviation training course/program titled

_____

Signed _____ Date _____

Cert. # _____ Expiration _____

I certify that _____
has satisfactorily completed the aviation training course/program titled

_____

Signed _____ Date _____

Cert. # _____ Expiration _____

I certify that _____
has satisfactorily completed the aviation training course/program titled

_____

Signed _____ Date _____

Cert. # _____ Expiration _____

TRAINING CERTIFICATIONS, CONTINUED

I certify that _____
has satisfactorily completed the aviation training course/program titled
_____

Signed _____ Date _____

Cert. # _____ Expiration _____

I certify that _____
has satisfactorily completed the aviation training course/program titled
_____

Signed _____ Date _____

Cert. # _____ Expiration _____

I certify that _____
has satisfactorily completed the aviation training course/program titled
_____

Signed _____ Date _____

Cert. # _____ Expiration _____

I certify that _____
has satisfactorily completed the aviation training course/program titled
_____

Signed _____ Date _____

Cert. # _____ Expiration _____

Additional Notes